Values and Objectivity in Science

Values and Objectivity in Science

The Current Controversy about Transgenic Crops

Hugh Lacey

LEXINGTON BOOKS

A Division of
ROWMAN & LITTLEFIELD PUBLISHERS, INC.
Lanham • Boulder • New York • Toronto • Oxford

LEXINGTON BOOKS

A division of Rowman & Littlefield Publishers, Inc.
A wholly owned subsidiary of The Rowman & Littlefield Publishing Group, Inc.
4501 Forbes Boulevard, Suite 200, Lanham, Maryland 20706

PO Box 317
Oxford
OX2 9RU, UK

Copyright © 2005 by Lexington Books

All rights reserved. No part of this publication may be reproduced, stored
in a retrieval system, or transmitted in any form or by any means, electronic,
mechanical, photocopying, recording, or otherwise, without the prior permission
of the publisher.

British Library Cataloguing in Publication Information Available

Library of Congress Cataloging-in-Publication Data

Lacey, Hugh.
 Values and objectivity in science and current controversy about transgenic crops /
Hugh Lacey.
   p. cm.
 Includes bibliographical references and index.
  ISBN 0-7391-1045-4 (cloth : alk. paper) — ISBN 0-7391-1141-8 (pbk. : alk. paper)
 1. Science—Philosophy. 2. Science—Social aspects. 3. Transgenic plants. I.
Title.

Q175.L158 2005
501—dc22
                                                    2004029939

Printed in the United States of America

∞™ The paper used in this publication meets the minimum requirements of American
National Standard for Information Sciences—Permanence of Paper for Printed Library
Materials, ANSI/NISO Z39.48-1992.

For Paige, Ashton, and Evelyn

Contents

Acknowledgments   xi

Introduction   1

PART I: THE INTERPLAY OF SCIENCE AND VALUES

1 How the Sciences Are and Are Not Value Free   17

    1.1 Values: the Modern Valuation of Control and its Presuppositions   18

    1.2 Science as Value Free: Impartiality, Neutrality, and Autonomy   23

    1.3 Links between Adopting Strategies and Holding Social Values—Materialist Strategies and the Modern Valuation of Control   28

2 Objectivity and Serving Human Well-Being   45

    2.1 Objectivity   45

    2.2 Science and Human Well-Being   54

3 The Distinction between Cognitive and Social Values   59

    3.1 Values, Value Judgments, and Value-Assessing Statements   60

    3.2 Social Ideals and Cognitive Ideals (The Aim of Science)   63

    3.3 Strategies, Methodological Rules, Metaphysics, and Social Values   67

    3.4 Three Moments of Scientific Activity   76

4 Incommensurability and "Multicultural Science"   81

    4.1 Competing Strategies   81

viii Contents

4.2	Comparing Theories Constructed under Competing Strategies	86
4.3	Choosing to Adopt a Strategy	87
4.4	"Multicultural Science"	92

5 The Social Location of Scientific Practices 97

5.1	Scientific Practices as Lacking Historicity	98
5.2	Kuhn's Account of the Historicity of Science	102
5.3	The Role of Applications	108
5.4	Agroecological Strategies	111
5.5	Adopting a Strategy and the Social Location of Scientific Research	117
5.6	Social Location and Adopting a Strategy: Additional Support for Pluralism	120

PART II: CURRENT CONTROVERSY ABOUT TRANSGENIC CROPS

6 The Controversy about Transgenics: Structure and Opposing Interests 125

6.1	Controversy about the Legitimacy of Using Transgenics	126
6.2	Suppositions of the Proponents' and Critics' Arguments	129
6.3	The Value-Outlooks of the Proponents and the Critics	132
6.4	Efficacy and Legitimacy	139

7 Strategies for Research in Agricultural Science 148

7.1	Seeds and the Knowledge They Embody	149
7.2	Seeds and Their Sociocultural Location	153

8 Benefits of Using Transgenics 165

8.1	Current Benefits	166
8.2	Anticipated Benefits	169
8.3	Golden Rice	171

9 Environmental Risks of the Development and Use of Transgenics 182

9.1	Risk: General Considerations	184
9.2	The Role of the Modern Valuation of Control	186
9.3	Disputes about Risks	189
9.4	Burden of Proof	192
9.5	Risks That Derive from the Socioeconomic Context of Use	198
9.6	The Logic of Endorsing Propositions about Risks	203

10	Alternative ("Better") Forms of Farming	211
	10.1 Agroecology: The Empirical Record	212
	10.2 Defending "Transgenics Are Necessary to Feed the World"	217
	10.3 The Challenge of Agroecology	219
	10.4 How to Conduct Scientific Research?	224
	10.5 Ethical Discourse and Scientific Inquiry	227
	10.6 Ground of Legitimating Current Uses of Transgenics: Science or Market Interests?	230

PART III: PROLEGOMENON TO EMPIRICAL INVESTIGATION OF FUTURE SOCIAL POSSIBILITIES

11	The Sociocultural Location of Alternatives to Transgenics	241
	11.1 Empirically Investigating " . . . Is Not Possible"	242
	11.2 Emancipatory Movements	245
	11.3 The Movements of the World Social Forum and Opposition to Neoliberalism	249
	11.4 A Question for Scientific Investigation: Is Another World Possible?	253

Appendix	261
Glossary of Abbreviations and Acronyms	273
Bibliography	275
About the Author	289

Acknowledgments

This book is a sequel to *Is Science Value Free?* (Lacey 1999/2004). It encapsulates and goes beyond work that I have published in articles since its publication. I wish to acknowledge permission, received from various publishers and editors, to use previously published material.

Chapters 1 and 2 represent a considerable expansion of Lacey (2002a): "The Ways in Which the Sciences Are and Are Not Value Free," in P. Gardenfors, K. Kijania-Placek, and J. Wolenski (eds.), *In the Scope of Logic, Methodology and Philosophy of Science*, volume 2, 513–26 (Dordrecht: Kluwer Academic Publishers); chapter 4 is a revised version of Lacey (2001): "Incommensurability and 'Multicultural Science,'" in P. Hoyningen-Huene and H. Sankey (eds.), *Incommensurability and Related Matters*, 225–39 (Dordrecht: Kluwer Academic Publishers). Material from these articles has been used with the kind permission of Springer Science and Business Media.

Chapter 3 is a revised version of Lacey (2004b): "Is There a Significant Distinction between Cognitive and Social Values?" in P. Machamer and G. Wolters (eds.), *Science, Values and Objectivity* (Pittsburgh: University of Pittsburgh Press). Used by permission of University of Pittsburgh Press.

Chapter 5 is a revised version of an article to be published in a volume edited by T. Rockmore and J. Margolis. Used by permission of the editor, Tom Rockmore.

Chapter 11 uses major parts of Lacey (2002b): "Explanatory Critique and Emancipatory Movements," *Journal of Critical Realism* 1:7–31. Used by permission of the editor, Mervyn Hartwig.

Chapters 6 through 10 (part II) consist largely of new and newly organized material that makes use (though none of these chapters corresponds directly to a particular article) of selected parts of several publications:

Lacey (2000a): "Seeds and the Knowledge They Embody," *Peace Review* 12: 563–69. Used by permission of Taylor & Francis (www.tandf.co.uk/journals).

Lacey (2002c): "Assessing the Value of Transgenic Crops," *Science and Engineering Ethics* 8:497–511. Used by permission of Opragen Publications.
Lacey (2002d): "Tecnociência e os valores do Forum Social Mundial," in I. M. Loureiro, M. E. Cevasco, and J. Corrêa Leite (eds.), *O Espírito do Porto Alegre* (São Paulo: Paz e Terra). Used by permission of the editor, Isabel Loureiro.
Lacey (2003a): "Seeds and Their Socio-cultural Nexus," in S. Harding and R. Figueroa (eds.), *Philosophical Explorations of Science, Technology and Diversity* (New York: Routledge). Used by premission of the editor, Sandra Harding.
Lacey (2004a): "Assessing the Environmental Risks of Transgenic Crops," *Transformação* 27: 111–31. Used by permission of the editor, Isabel Loureiro.

I acknowledge, with thanks, receiving financial support from the following organizations while preparing this book:

NSF: National Science Foundation (SES 0322805), which provided a grant to support my proposal, "The Structure of the Ethical Controversy about Transgenics." [The opinions expressed in this book are my own and do not necessarily represent those of NSF.]
FAPESP: Research Foundation of the State of São Paulo, Brazil.
Swarthmore College: Faculty Research Fund, and Scheuer Family Professorship of the Humanities.

In recent years I have taught courses on the material of this book on several occasions: courses at Swarthmore College (2001, 2002), University of Pennsylvania (2002, 2004), Universidade de São Paulo (2000, 2004); and "mini-courses" at Universidade Federal de Santa Catarina (2000), Universidade Federal de Bahia (2003), and UNISINOS, Universidade do Vale do Rio Sinos (2004). I have benefitted from interactions with students on all these occasions. I also benefitted from discussions after lectures I gave at the NEH Summer Philosophy Institute, "Science and Values," at University of Pittsburgh (2003), directed by Peter Machamer and Sandra Mitchell (I am grateful to them for the invitation to participate); and also (especially with respect to chapter 6) from discussion after papers I presented at the conferences: "A questão das plantas transgênicas: verso e anverso," Centro de Estudos da Saúde do Trabalhador e Ecologia Humana (CESTEH) na escola Nacional de Saúde Pública da Fundação Oswaldo Cruz, Rio de Janeiro (June 2004), organized by Carmem Marinho, and "Transgenic crops: benefits, risks and alternatives," 4S-EASST Conference, *Public Proofs: Science, technology and democracy*, Ecole des Mines de Paris (August 2004).

Many colleagues have contributed to the development of the ideas presented in this book, and I am grateful to them all. Richard Eldridge (Swarthmore) and

Marcos Barbosa de Oliveira (São Paulo) have read and made helpful comments on almost every chapter and supplied me with information on various topics. I also wish to thank especially:

Amy Vollmer, professor of biology, Swarthmore College, with whom on three occasions I cotaught a course, "Biotechnology and Society: The Case of Agriculture," and from whom I learned an enormous amount about biotechnology and molecular biology in general and transgenic technology in particular.

Miguel Guerra and Rubens Nodari, professors of agronomy, Universidade Federal de Santa Catarina, Florianópolis, Santa Catarina, Brazil, in several conversations and numerous emails, have taught me a great deal both about alternatives to transgenic-oriented agriculture that are being developed in Brazil (and research that informs it) and new dimensions of the critique of transgenic technology, and they have also responded to some of my papers in critical and helpful ways.

Pablo Mariconda and Mauricio Carvalho de Ramos, São Paulo, have been wonderful collaborators in developing the ideas of part II (and they organized and participated in various forums in which I was able to test some of these ideas).

Paul Thompson, professor of philosophy and agricultural science, Michigan State University, has shared with me, in a few conversations, his enormous knowledge about various forms of farming and, in particular, attuned me to the merits of a nuanced pro-transgenics position.

Marcelo Leite, science editor, *Folha de São Paulo*, has provided me regularly with material pertaining to the ongoing controversy in Brazil, including at the legislative level.

Introduction

This book is a sequel to *Is Science Value Free? Values and Scientific Understanding* (Lacey 1999/2004; hereafter referred to as *SVF*). I introduced in *SVF* a general model of the interplay of science and values that enables us to identify clearly the ways in which values legitimately play a role in scientific practices and the ways in which they do not. I was interested, however, not only in the impact of values on scientific methodology, but also (though this was much less developed) in how scientific practices and results may have impact in the realm of values, in how, for example, science may have implications for and contribute to the quest for social justice and human well-being. In this present book, the question of how to conduct scientific practices for the sake of furthering human well-being informs the overall argument.

Part I does deal with questions of scientific methodology. It strengthens some of the arguments made in *SVF*, draws new consequences that flow from the general framework concerning how scientific results are to be interpreted, argues for the methodological importance of applied science, and deepens arguments in favor of methodological pluralism by bringing to the forefront questions that must be addressed in order to legitimate applications of scientific knowledge—all so as to show that the quest for objectivity in science is not incompatible with conducting science for the sake of furthering human well-being. Then, in part II, I bring the analysis of methodology developed in part I to bear on a current controversy with far-reaching ethical and social implications: whether transgenic crops should be used widely and, if not, what forms of farming provide suitable alternatives to using them. Through drawing on this analysis, we gain a rich understanding of the scientific, ethical, and political dimensions of the controversy and of how the dimensions interact with one another. At the same time, all the themes of part I are illustrated, especially as they concern the conduct and institutionalization of scientific practices. Throughout part I, I anticipate various features of the discussion to follow in part II, introducing them where convenient from chapter to chapter, in order to illustrate, to stimulate, and to give concrete texture to the general philosophy of science being developed; and also in order to be in a position to make soundly based recommendations about how to respond to the controversy.

The analysis of the transgenics controversy, in the light of my general model of the interplay of science and values, is the most distinctive feature of this book, and it is intended to have impact on public discussion. But the argument of the book will not be dated when the controversy no longer captures attention, for that is not what the book principally is about. It is about my model of the interaction of values and scientific practices, and about making sense of science as a historical and social phenomenon. It is a test of the explanatory power and significance of my model that it makes sense of the actual historical outcome of the transgenics controversy, whatever it may be. Another, ultimately more important, test is whether the analysis can be generalized to make sense of a wide range of scientific developments and controversies, and whether some of the alternative kinds of scientific practices to which the analysis points can be developed into flourishing areas of research and practice.

SUMMARY OF THE CONCLUSIONS OF *IS SCIENCE VALUE FREE?*

The principal theme of *SVF*, as its title makes apparent, is the analysis and critical appraisal of the widely held view that science is value free. In order to locate the contribution of this book, it will be helpful to recall the central theoretical concepts and to state the most important conclusions of the earlier book. I will do this by presenting a list of the principal theses that were defended in it.

> Thesis 1: The idea that science is value free is best understood as a combination of claims about three key aspects of scientific practices—the acceptance of theories and the knowledge claims that are represented in them, the consequences of applying scientific knowledge, and scientific methodology—that I call respectively: impartiality, neutrality, and autonomy; it is well captured by the thesis: impartiality, neutrality, and autonomy are constitutive values of scientific practices and institutions—where, among other things, impartiality presupposes that there is a distinction between cognitive and social (ethical and other kinds of non-cognitive) values, and neutrality presupposes that scientific theories have no value judgments among their logical implications.

This thesis, and the presuppositions of impartiality, neutrality, and autonomy, will be re-articulated in a sharper and more elegant manner in chapter 1 of this book.

After extensive argument, I concluded:

> Thesis 2: Only impartiality can be unambiguously upheld. It expresses the value: to accept a theory of a domain of phenomena if and only if it manifests the cognitive values to a suitably high degree in the light of relevant available empirical data, and to reject a theory if and only if it is inconsistent with a soundly accepted theory; hence there is no proper role for ethical and social values, alongside the cognitive values, in making judgments of theory acceptance. In contrast, autonomy is not a realizable value; and neutrality—that, overall and in principle, the application of scientific knowledge serves value-outlooks evenhandedly—is compromised within mainstream scientific practices, but could be more fully manifested if scientific research were conducted under a suitable plurality of methodological approaches.

The key to my argument lay in introducing, as an element in the analysis of scientific methodology, in addition to "theory" (hypothesis) and "empirical data," what I call a "strategy":

> Thesis 3: (a) Scientific research is always conducted under a strategy, whose main roles are, first, to prescribe constraints on the kinds of theories (and the kinds of categories they may deploy) that may be entertained and investigated, and thus to specify the kinds of possibilities that may be identified in the course of the research, and, second, to select the relevant kinds of empirical data to seek out and report, and the phenomena and aspects of them that are to be observed and experimented upon. (b) The aim of science permits that successful research may be conducted under a variety of kinds of strategies.

But:

> Thesis 4: Modern scientific research has been conducted almost exclusively under particular kinds, of strategies (albeit a considerable variety of them), those I call "materialist strategies," under which theories are constrained to those that represent phenomena and encapsulate possibilities in terms of their being generable from underlying structure (and its components), process, interaction, and the laws (characteristically expressed mathematically) that govern them; and, by virtue of obtaining them as products of measurement, instrumental, and experimental operations, empirical data are generally quantitative.

Representing phenomena under materialist strategies decontextualizes them, by dissociating them from any place they may have in relation to social arrangements, human lives and experience, from any link with value, and from whatever social, human, and ecological frameworks in which they may be embedded. In this book, I call the kinds of possibilities that can be encapsulated under materialist strategies "decontextualized possibilities" (or, as I called them in *SVF*, "material possibilities"). In order to illustrate that not all possibilities that may be identified in systematic empirical inquiry (whose results accord with impartiality) are reducible to decontextualized possibilities, I used the examples of human agency and agroecology. Research conducted on human and agroecological phenomena under respectively "feminist strategies" (*SVF*, chapter 9) and "agroecological strategies" (chapter 8) has produced knowledge in accordance with impartiality, but under these strategies phenomena are not dissociated from their human/social/ecological contexts, and so the possibilities that are identified for them are not decontextualized.

> Thesis 5: Scientific research—systematic empirical inquiry that produces results that are in accordance with impartiality—may be conducted (for some domains of phenomena) under strategies that, while they may freely utilize results consolidated under materialist strategies, are not reducible to materialist strategies.

Thesis 5, backed by the illustrations, confirms that including (b) in Thesis 3 is not merely an abstract logical point.

Theses 4 and 5, then, lead us to pose the questions: How do we explain the fact that modern scientific research has been conducted almost exclusively under materialist strategies? And: Are there good reasons for conducting research in this way? After considering and rejecting a variety of answers that have been put forward, I concluded:

> Thesis 6: (a) The almost exclusive adoption of materialist strategies in modern science is explained (i) by their fruitfulness and potential for practically unlimited further development, (ii) by the fact that there are mutually reinforcing relations between adopting them and holding a set of social values, specifically the modern valuation of control, and (iii) by the fact that the modern valuation of control is widely upheld throughout advanced industrial countries and highly embodied in their leading institutions. (b) There are good reasons for the privilege that materialist strategies have gained only to the extent that there are good reasons to uphold the modern valuation of control.

The modern valuation of control refers to a set of specifically modern values connected with the control of natural objects, having to do with expanding the scope of technological control, its value not being systematically subordinated to that of other ethical and social values, and the degree of its penetration into modern lives, experience, and institutions. It and its presuppositions are discussed more fully in section 1.1. It does not follow from this thesis that materialist strategies are always adopted because of an interest to further the modern valuation of control, or that movement from one kind of materialist strategies to another can be explained by reference to these social values. Thesis 6 concerns the almost exclusive adoption of materialist strategies in modern science, and it relates this to particular social values being widely upheld. Item (b) is crucial.

Where the modern valuation of control is contested, there can be no objection (in principle) to adopting strategies in research in virtue of their mutually reinforcing relations with other values (subject, of course, to providing reasons to uphold these values). Then, we can see Thesis 6 to be a particular case of the more general:

> Thesis 7: Social values may provide a compelling reason to adopt a particular kind of strategy: adopt strategies in view of mutually reinforcing relations that adopting them may have with holding specific social values. In practice, this may mean: adopt strategies under which valued kinds of possibilities (if there are any) can be systematically identified and the means for realizing them discovered, or that have the potential to produce results that, on application, can further the interests defined by the values—subject always to the conditions of (i) fruitfulness, (ii) the results gained being in accord with impartiality, and (iii) the recognition that it is not evidence against the genuineness of a possibility that it cannot be identified under a favored strategy.

Feminist values may provide a good reason to adopt "feminist strategies," and the values of "popular participation," widely held values within movements of small-scale farmers and rural workers in many of the impoverished regions of the world, may provide a good reason to adopt "agroecological strategies" (see section 5.4 and part II of this book). Thesis 7 goes hand in hand with:

> Thesis 8: The moment of deciding to adopt a strategy may be logically separated from that of choice to accept or reject a theory (of a specified domain of phenomena) constructed under the strategy, so much so that commitment to impartiality can be maintained at the latter moment, even though social values may have a legitimate role at the first moment. Moreover, the social values in play at the first moment may be the same values whose furtherance is served at a third moment, that of the application of scientific knowledge.

Theses 7 and 8 together sum up the general model of the interplay of science and values that I referred to at the outset.

Research conducted under one kind of strategy may complement that conducted under another by, for example, exploring possibilities of things that cannot be considered because of the constraints of the other. But strategies may also compete (e.g., for resources), and this may make it socially impossible for research to be conducted simultaneously and in a probing way under conflicting strategies. Thus, if one kind of strategies is privileged because of its links with predominant social values, this may lead to inability even to recognize that there is a choice of strategy to make. Specifically:

> Thesis 9: So strong is the grip of materialist strategies in modern science that it is often not appreciated that there may be certain domains of phenomena (e.g.,

agriculture), which are of special salience where the modern valuation of control is contested, but whose possibilities cannot be adequately encapsulated in theories confirmed in research conducted under materialist strategies, although they can be under other kinds of strategies (e.g., agroecological strategies).

With this in mind, I introduced:

> Thesis 10: The aim of science is best served by institutionalizing scientific practices so that a plurality of strategies, linked respectively with different social values, may be actively pursued. This would also make possible the fuller manifestation of neutrality and giving better attention to value issues raised by applications, and, above all, be conducive to strengthening the institutions of democratic participation.

THE PLAN OF THIS BOOK

Part I

These theses all remain intact in this book, and I will make use of them frequently. In chapter 1, I will recapitulate the arguments in favor of them, clarify them in various ways, emphasize how they are illustrated by the competition between agroecological and materialist strategies, and anticipate the conclusions of later chapters. The other chapters of part I provide arguments complementing those of *SVF* and draw further implications from the theses concerning both how to understand and how to conduct scientific practices. In chapter 2, I will show that there need not be conflict between the traditional ideal of scientific objectivity and the conduct of scientific inquiry for the sake of furthering social justice and human well-being. In chapter 3, I address a gap in the argument of *SVF*. My defense of impartiality, which draws on the presupposition that there is a significant distinction between cognitive and social (ethical and other kinds of) values, has been criticized. In this chapter I offer a sustained defense of the importance of the distinction. In the course of doing so, new insights arise about important methodological issues connected with applications of scientific knowledge. Distinguishing between the efficacy and legitimacy of proposed applications, and reinforcing the importance of Thesis 10, I show that legitimacy depends on endorsing hypotheses (e.g., about risks to human health and the environment, and about the availability of alternative efficacious means to realize the objective of the application) that lie beyond the purview of research conducted under materialist strategies. The significance of this conclusion is made apparent throughout part II.

In chapter 4 and chapter 5 I explore more fully the implications of the pluralism that I have proposed. What is the range and variety of strategies that could or should be developed? I do not attempt to answer this question fully; to do so would require a vast number of case studies in fields like psychology, medicine, and energy policy. I do suggest in chapter 4, however, drawing upon the case study of agriculture, that a wide range of cultural values may legitimately have impact on the kinds of strategies that one adopts, so that there are legitimate culture-based variations in approaches to scientific practices. Then, in chapter 5, developing this argument and making use of a detailed account of agroecological strategies, I consolidate the conclusion that there are rich dialectical links between methodology and application, so much so that it is often impossible to separate the interpretation of scientific results from the social location in which the research is conducted. From this, it follows that scientific practices exhibit historicity: that their character changes, and must change, in fundamental ways that arise historically, through being responsive to and shaped significantly by historical and cultural variations in the realm of daily life and experience and in the structures of social practice.

The arguments I make in part I—for strategic pluralism, for there being (often and legitimately) mutually reinforcing relations between adopting strategies and holding particular social values, and for the historicity and sociocultural shaping of scientific practices—are arguments in the philosophy of science. They draw principally upon my statement of the aim of science (section 3.2) and my exploration of how to further the manifestation of the widely acclaimed scientific ideals of impartiality and neutrality. In order to show that this plurality represents more than an abstract possibility (Thesis 5), I introduced the case of agroecological strategies (detailed in section 5.4) as a concrete illustration. Agroecological strategies are not reducible to materialist strategies, and adopting them has mutually reinforcing relations with holding the values of popular participation (characterized in section 6.3). The soundness of the argument in part I does not depend on holding any particular ethical/social values (apart from those implicit in the aim of science). My highlighting of agroecological strategies does reflect my own commitment to the values of popular participation. This commitment, however, is irrelevant to the appraisal of the fruitfulness of these strategies; and also the facts that they are not reducible to materialist strategies, and that these values contest the modern valuation of control, are irrelevant to this appraisal.

In part I, the role of agroecological strategies is to provide an example that shows that there are actual instances of what philosophical analysis identifies to be possible (a plurality of fruitful strategies). Other strategies, for example, in the psychological, social, or medical sciences, could have played this role just as well. In part II, the role of agroecological strategies is essential; knowledge gained in investigations conducted under them is indispensable for making important ethical judgments about the legitimacy of using transgenics; and development of the farming practices that agroecological knowledge informs is important for the consolidation of democratic ideals.

Part II

In part II, drawing on the conclusions of part I, I offer an interpretation of current controversy about transgenic crops and alternative types of farming such as agroecology. The controversy is about the legitimacy of research, development, practical agricultural implementation of transgenics, and practices and policies (pertaining to transgenics) that currently are being implemented under the sponsorship principally of agribusiness corporations. I take the pro-transgenics side to argue for the legitimacy (and importance) of the development, immediate implementation, intensive utilization, and widespread diffusion of transgenics in the agricultural practices that produce major crops, throughout the world as soon as possible, and for support for transgenics to become a central plank in national and international agricultural policies. And I take the con side to deny that the pro conclusions have been adequately established; to maintain that more research is needed before a definitive position can be taken; and positively, to prioritize alternatives that do not use transgenics, such as agroecology, and the urgency and priority of investigating their productive potentials.

My interpretation identifies the principal points of contention (while recognizing, since there is a variety of opinions in play on both sides, that it involves a certain amount of idealization). This is a prerequisite to exploring what would have to be done (if anything can be) to bring about—or to show that there are insuperable obstacles to bringing about—a resolution of the dispute. It involves two steps: first, identifying four pairs of contrary propositions that are in dispute (section 6.2)—pro: P_1–P_4 and con: C_1–C_4; and, second, sketching the value-outlooks that are implicated, respectively, in the two positions (section 6.3). The propositions are about strategies for research in agricultural science (discussed in detail in chapter 7), benefits (chapter 8) and risks of using transgenics (chapter 9), and whether there are better alternatives (chapter 10):

Strategies for Research in Agricultural Science

P_1 Developments of transgenics are informed in an exemplary way by scientific knowledge, that is, they are informed by knowledge gained in research conducted under appropriate versions (biotechnological) of materialist strategies; they are instances of techno-scientific developments, which are the principal sources of improvements of agricultural practices and (more generally) meeting human needs.

C_1 The kind of knowledge gained under materialist strategies is incomplete and cannot encompass the possibilities of, for example, sustainable agroecosystems and the possible effects of uses of transgenics on the environment, people, and social arrangements; it is necessary to adopt other strategies in order to investigate these matters.

Benefits of Using Transgenics

P_2 There are great benefits to be had from using TGs now, and these benefits will greatly expand with future developments, among which are promised TG crops with enhanced nutritional qualities that can readily be grown in poor developing countries so that TGs may become key to addressing problems like those of hunger and malnutrition. When these promises are fulfilled, the benefits of TGs will become spread evenhandedly so as (in principle) to serve the interests and to improve the farming practices of groups holding any viable value-outlooks.

C_2 The benefits claimed for currently used TGs reflect the ethical/social values of agribusiness, large-scale farmers, and others who are beneficiaries of the global market. Furthermore, not only are the benefits relatively slight (perhaps even exaggerated by the proponents), being confined largely to these groups and not extending to small-scale farmers in the "developing" world (or to organic farmers in the advanced industrial societies), but also the promises made about future benefits are not credible, in part because developments of TGs reflect the interests of the global-market system, the very same system within which poverty, the fundamental cause of hunger and malnutrition, persists today.

Risks of the Development and Use of Transgenics

P_3 There are no hazards to human health or the environment arising from the current and anticipated uses of transgenic crops and their products that pose risks—of seriousness, magnitude, and probability of occurrence sufficient to cancel the alleged value of their benefits—that cannot be adequately managed under responsibly designed regulations.

C_3 This claim about risks is not well established scientifically. Moreover, the greatest risks may not be direct ones to human health and the environment mediated by biological mechanisms, but those occasioned by the socio-economic context of the research and development of transgenics and their associated mechanisms, such as designating that transgenic seeds are objects to which intellectual property rights may be granted.

Alternative (or "Better") Forms of Farming

P_4 There are no alternative kinds of farming that could be deployed instead of the proposed transgenic-oriented ways without occasioning unacceptable risks (e.g., not producing enough food to feed and nourish the world's growing population), and that reasonably could be expected to produce greater benefits concerning productivity, sustainability, and meeting human needs— "transgenics are necessary to feed the world."

C_4 Agroecological methods (and other alternatives) can be and are being developed that enable high productivity of essential crops (and occasion relatively less risk); and they promote sustainable agroecosystems, utilize and protect biodiversity, and contribute to the social emancipation of poor communities. Furthermore, there is good evidence that they are particularly well suited to ensure that rural populations in "developing" countries are well fed and nourished, so that without their further development current patterns of hunger are likely to continue.

By identifying these as the key points of contention, I hope to have interpreted the dispute so that a perspicuous contrast is made between the two sides, one that meets the following conditions: (i) each side can acknowledge that its position has been fairly represented; (ii) each side is enabled to recognize the internal coherence of the other, to identify clearly what lies behind the disagreements, and to raise questions about the evidence and arguments that support the various propositions; (iii) avenues that might lead to resolution, which are in continuity with the basic commitments of each side, become opened for exploration.

Implications of the Interpretation

Although I think that my interpretation meets these conditions, this does not mean that I abstain from taking positions on the propositions. There is good reason—I will argue, again with grounding in the conclusions of part I—to endorse C_1 (methodological pluralism) (chapter 7), that now P_3 and P_4 lack the support that they need in order to play their role in arguments legitimating uses of transgenics (chapter 9 and chapter 10, respectively), and that there is urgency to conduct research relevant to test the limits of the promise of alternative agricultural methods, expressed in C_4 (chapter 10). That is enough to deny legitimacy at the present time to projects aimed at the widespread implementation of transgenic-oriented agriculture throughout the world. But it also is part of my argument that the legitimacy of the transgenics project in the long run depends on the outcomes of testing the limits of C_4 (section 10.4), so that using my interpretive framework does not guarantee that the opponents of transgenics will be vindicated in the long run.

The interpretive framework sets up a context in which empirical investigation, conducted under a plurality of strategies (including agroecological ones), could play a major role in cutting through the disagreements about risks (P_3/C_3) and alternative types of farming (P_4/C_4). Conducting this kind of research would inform ethical deliberation in two ways, by providing knowledge (a) for appraising presuppositions of the legitimacy of using transgenics on a wide scale at the present time, and (b) for informing agroecological innovations that are important for bringing about greater manifestation of the values of popular participation. Thus, engaging in research conducted under agroecological strategies is likely, because of (a), to further the manifestation of impartiality and, because of (b), of neutrality; and so, adopting strategies, which have mutually reinforcing relations with the values of popular participation, is likely to contribute to the furtherance of acclaimed scientific interests. According to my interpretation, the con argument is not an abstract one, and it does not involve merely negative criticism of mainstream science, since it is also rooted in critical reflection on the practices of agroecology. It enables a positive case to be made for the scientific significance of the knowledge that informs agroecological practices and for the value of research that strengthens them. It is part of a philosophical perspective that interprets and supports both the practices and the research conducted to inform them (as having a proper place—alongside others—within scientific practices), and it defends their credentials from criticisms that they are "unscientific." So it is an interpretation that confronts the predominant self-image of contemporary science with the sound claims of an alternative practice. The strength of the argument goes hand in hand with the value and viability of the alternative practices.

Does the Authority of Science Provide Backing for the Pro Transgenics Side?

I said that I wanted an interpretation in which each side would acknowledge the portrayal made of it. I have taken seriously the pro side's claim to have the backing of science. Obviously developments of transgenics are products of research conducted under materialist (biotechnological, molecular biological) strategies, and their efficacy (within certain domains) has been confirmed by this research. In addition, I take the pro side to claim scientific backing for the key propositions about risks and alternatives that are important for legitimating uses of transgenics. I argue (chapter 9; chapter 10) that at the present time there is not strong empirical backing for P_3 and P_4, since relevant inquiries (that I specify)—requiring the use of a plurality of strategies—pertaining to risk have not been conducted, and others pertaining to alternatives have effectively been ignored. But, by endorsing P_1, the pro side tends to identify scientific research with research conducted under materialist strategies, and so it does not recognize the possibility (and, in this case, necessity) of scientific research conducted under a plurality of strategies. (Thus, it tends to interpret the con side as "unscientific" or even "antiscientific.") I will suggest that endorsing P_1 is a consequence of holding the modern valuation of control and endorsing its presuppositions (section 9.2); then the absence of strong empirical backing for P_3 and P_4 (and accepting that there is a strong presumption in their favor) derives, not from scientific evidence, but in part from a value commitment, that has presuppositions (e.g., that techno-scientific solutions can be found for virtually all socially significant problems, and that there are no significant possibilities for value-outlooks, not incorporating the modern valuation of control, to be actualized in the foreseeable future—section 1.1) that themselves cannot be investigated under materialist strategies.

At the same time as the pro side claims the backing of the authority of science, it also represents the interests of leading institutions of capital and the market that dominate the world economy today. Another way to look at its endorsement of P_1, and its endorsement of the presuppositions of the modern valuation of control, is that they derive from endorsing that there are no significant (valued) possibilities for the foreseeable future outside of the trajectory of the institutions of capital and the market. Then (section 10.6), the pro argument could be strengthened, since it would appear to marginalize the relevance of C_3 and C_4, by replacing P_4 with P_4a:

P_4a: There are no alternative kinds of farming—*within the trajectory of the socioeconomic system based on capital and the market*—that could be deployed instead of the proposed TG-oriented ways without occasioning unacceptable risks (e.g., not producing enough food to feed and nourish the world's growing population), and that reasonably could be expected to produce greater benefits concerning productivity, sustainability, and meeting human needs; *and outside of this trajectory there are no genuinely realizable agricultural possibilities*.

P_4a may be taken to express a political-economic commitment and, given the constraints it states, empirical research might contribute to vindicate it; and, given the economic and political power linked with it, it might be expedient simply to dismiss the con side as a nuisance. Then the authority of science would be subordinated to the political values and power embodied in this trajectory. Alternatively, the pro side might claim to endorse P_4a on empirical grounds and, in this way, to reclaim the authority of science. Clearly the con side would oppose it on both counts.

Part III

In the public debates about transgenics, the pro side often moves imperceptibly back and forth between P_4 and P_4a. Responding to P_4, the con side emphasizes the fruitfulness of research conducted under agroecological strategies and the promise of agroecological approaches to farming; responding to P_4a, it affirms the viability of the movements that embody the values of popular participation and their potential to grow with a trajectory that could nurture new kinds of social structures. Thus, for the con side, the development of research conducted under agroecological strategies, the development and improvement of agroecological farming, and the activities and growth of movements that embody the values of popular participation are inseparably linked.

These considerations all raise the question of how propositions like P_4a, the presuppositions of the modern valuation of control, and other questions about future social possibilities—as well as the various contrary propositions that would be affirmed by the con side—can be investigated in a systematic empirical way. Under what kinds of (social science) strategies would the investigations have to be conducted? The answer to this question, and the outcomes of the research, are relevant to attempts to resolve the controversies about transgenics in ways that make use of the input of scientific (systematic empirical) investigation to the utmost. Those with power on their side have not waited for an answer before going ahead with the rapid and widespread introduction of transgenics; consequently the con side often finds itself in a negative reactive mode, opposing what is happening. That should not obscure the continuing importance of investigating further the promise contained in C_4. Evidence for C_4 directly challenges the empirical credentials of P_4, but it also remains important that there be movements that challenge P_4a. The possibility of manifesting values like those of popular participation to a greater degree depends on there being genuine alternative practices (in agriculture as well as other areas) that may reflect these values, and the latter depends on social and political action that claims and gains more and expanding spaces for these practices, expansion that is not possible without successfully challenging P_4a.

In the brief part III, I begin to entertain questions about how to investigate future social possibilities. I do just enough to make clear that the conclusions of part I, especially those about strategic pluralism and mutually reinforcing relations between adopting strategies and holding values, will play an important role in answering them. I can do no more than this within the scope of this book.

Part I

THE INTERPLAY OF SCIENCE AND VALUES

Chapter One

How the Sciences Are and Are Not Value Free

The spectrum of commonly stated aims of science may be identified by its extremes: understanding and utility. Understanding involves the description (and thus classification) of phenomena, their explanation, and encapsulation of their possibilities—answers to: What? Why? What is possible? and often also to How? (section 3.2) Scientific understanding is expressed in theories and empirically grounded, and the criteria for its appraisal will be called cognitive values. The cognitive values include, among others, empirical adequacy, explanatory power and intertheoretic consistency, features that may be manifested in theories to a greater or lesser degree. Cognitive values are the features whose high manifestation is desired of acceptable theories, that is, theories in which sound understanding of a domain of phenomena is represented. Modern science has been an unstoppable font of sound understanding, which in turn has been widely, effectively, and usefully applied. Part of the explanation usually offered for this twofold success draws upon the view that science is value free, a view that is best treated as conjoining claims about three key aspects of scientific practices: (i) the acceptance of theories and the knowledge claims that are represented in them, (ii) the consequences of applying scientific knowledge, and (iii) scientific methodology; I call the claims, respectively "impartiality," "neutrality" and "autonomy." In this chapter I will summarily characterize each of these three claims, and sketch the main lines of my critical appraisal of them. In doing so, I will recapitulate the argument of *SVF* (summarized in the introduction), once again highlighting the central methodological role of a strategy. I will also anticipate the conclusions to be reached in the chapters that follow, and begin to illustrate them using the example of agricultural research, which will become the focal point of part II. First, however, I introduce some brief remarks about values.

1.1 VALUES: THE MODERN VALUATION OF CONTROL AND ITS PRESUPPOSITIONS

Any articulation and appraisal of the view that science is value free reflects a conception of the nature of values. I will deploy the account of values that I have developed in *SVF* (see also section 3.1; section 6.3; chapter 11 for more details). Unlike in noncognitive and subjectivist accounts, according to it, values of various kinds (personal, moral, social, etc.) are held together, mutually reinforcing one another, in value-outlooks, rendered coherent, ordered, and rationally worthy of being held by certain presuppositions about human nature (and nature) and about what is possible, presuppositions that are open in some measure of empirical investigation. It follows, then, that the outcomes of scientific inquiry may contribute to support or undermine the rational credentials of a value-outlook. I will say that a value-outlook is viable provided that its presuppositions are consistent with soundly accepted scientific knowledge. Viability is a necessary condition for the rational adoption of a value-outlook. Not all value-outlooks that have been entertained are viable, but the advance of science leaves open a range of viable ones. That is why it is coherent to see scientific developments as having played a rational role in the demise of the value-outlook of medieval Christendom, while also maintaining that science is impotent to adjudicate the great value disputes of our age.

It is often assumed that any value-outlook rationally held today must include (what I call) "the modern valuation of control," a set of values connected with the control of natural objects and techno-scientific advances, where "techno-science" refers to the combination of technological advances informed by scientific knowledge gained in mainstream scientific investigations (i.e., in research conducted under materialist strategies) and of the research and development that leads to practical implementation of its results for controlling natural objects and processes (technology). Thus, for example, those who cite the value of organic farming over that of the agricultural practices spawned by recent innovations of biotechnology tend to be dismissed on the ground that they run counter to the trajectory set by the modern valuation of control. But that dismissal does not follow from currently soundly accepted scientific knowledge. That a value-outlook rests uneasily with the current centers of power does not imply that adopting it violates canons of rationality, or that it is not viable.

The Modern Valuation of Control

Matters pertaining to the modern valuation of control will play a key role in the criticisms of neutrality and autonomy that I will make, so I will dwell on it a little and, at the same time, illustrate my contention that value-outlooks are rendered rationally acceptable in light of presuppositions that serve to make them coherent. While valuing some measure of control of natural objects is surely a human universal, attitudes held towards it in modernity are quite distinctive both with respect to the extent of its reach and to its mode of relationship with other values (*SVF*: 111–15). I identify the components of the modern valuation of control as follows:

- The instrumental value of natural objects is dissociated from other forms of value; then the exercise of control over natural objects becomes per se a social value that is not systematically and generally subordinated to other social values.
- Expanding human capabilities to control natural objects, the widespread institutional embodiment of these capabilities, and especially creating new technologies—"progress"—rank high as values.
- Control is the characteristic human stance to adopt toward natural objects. Exercising control and, above all, engaging in the research and development projects in techno-science in which our powers to control are expanded, are essential and primary ways in which we express ourselves as modern human beings, in which are cultivated such personal "virtues" as creativity, inventiveness, initiative, boldness in the face of risks, autonomy, rationality, and practicality. Thus, an environment which is shaped so that many and varied possibilities of control may be routinely actualized in the course of daily life, one dominated by techno-scientific objects, is highly valued, as is the spread of technology into more and more domains of life and the definition of problems in terms of their having a technological solution.
- The values that may be manifested in social arrangements are, to a significant extent, subordinate to the value of implementing novel techno-scientific advances, which have prima facie legitimacy so that a measure of social disruption may be tolerated for their sake, and whose side effects may be addressed largely as "second thoughts."

The last item is reflected in the fact that practices for the systematic empirical investigation of "side effects," especially social and ecological ones, are underdeveloped compared to those that deal with novel possibilities of control (section 3.4; section 6.4; chapter 9). This, in turn, reflects a negative valuation of "intrusive" governmental regulation of techno-scientific innovation. (One might add this as another component of the modern valuation of control.)

As articulated here, the modern valuation of control concerns control exercised on material and biological objects; it is silent about exercising control over human beings. It involves valuing techno-science and its practical applications highly. Nevertheless, since how we interact with material objects is intricately connected with how we interact with other human beings, when techno-science dominates the productive process, the control of material objects may de facto require practices in which human behavior is also subject to control (Lacey 1990). This creates permanent tensions in the human condition, for people may aspire to exercise control over material objects only to the extent that it does not imply control over human beings, and to introduce controls for the sake of human autonomy and choice, not for the domination of other human beings. It is not at all clear that the modern valuation of control can be manifested to a high degree without being implicated in the domination of (some) human beings and in considerable social and ecological devastation (section 11.2). Where the modern valuation of control is held, since such questions arise only as "second thoughts," they might not be seen to be of pressing concern. They are hardly seen as matters that should curtail the unfolding of "progress," so that holding these values firmly can induce a measure of tolerance to exercising control over human beings. This is reinforced by some of the presuppositions (below), especially Presupposition (e) when it is expanded into full-blown materialist metaphysics. Then, human beings are thought of in terms of the same categories (underlying structure, process, interaction, and law) as material objects are, and so our stance toward them may become essentially the same (control).

The modern valuation of control stands in stark contrast to many other cultural outlooks in virtue of its not subordinating control of natural objects systematically to other social values. Roughly, "control" suggests effectively using things (or people) for one's own ends, doing to them what one wills, manipulating or coercing them, exercising power over them, treating them as objects of value for oneself, and subordinating any other value they have to this. In the case of people, control contrasts with relations such as mutual participation, dialogue, cooperation, collaboration, being in a community (with shared goals and different roles in cooperating to achieve them); with mutual acknowledgment, recognition and respect, with mutual empowerment, with relations that are based on and that nurture reason, with persuasion rather than manipulation. In the case of natural objects, it contrasts with relations of attunement and adaptation, stewardship or facilitation (management) that balances human needs and well-being with nurturing and restoring the natural environment (and agroecosystems). Control emphasizes one's goals (considered as defined in relation to one's personal values) and one's efficacy in realizing them. The contrasting stances emphasize what is to be brought into balance in forming one's goals, subordinating efficacy to maintaining (or restoring) human, social, and ecological well-being, so that from their perspective the distinction between "immediate effect" and "side effects" of controlling interventions cannot be clearly drawn. There is not, for example, a distinction between "production of a crop" and the side effect of "ecological devastation"; they are just different aspects of one and the same activity (section 5.4).

Control is universally valued. Those who contest the modern valuation of control do not deny this; rather they wish to subordinate the value of control to that of sustaining an appropriate balance of instrumental with other forms of value. This is a balance that derives from the sense that natural objects have their own integrity and value in virtue of their place in ecological and/or cosmic systems ("wholes"), and where control is balanced by general patterns of renewal, nurturing, cultivation, attunement, love, stewardship, restoration, mutuality—analogues of "dialogue" between humans (cf. Keller 1996; Maxwell 1984; Tiles 1996; often, see section 5.4; chapter 7, these viewpoints challenge Presupposition (e) below.) Such outlooks persist and retain important roles today, for example, in art and music, and—as we will see in part II—among those who oppose the expansion of techno-science into agricultural practices and the members of "popular organizations" in many of the impoverished regions of the world. The modern valuation of control is most dominant in productive and consumer domains of modern life, though it tends to expand into ever-more domains of life.

Presuppositions of the Modern Valuation of Control

The modern valuation of control is highly manifested throughout the world today, in part because it is upheld by and highly embodied in predominant economic and political (and military) institutions, so that holding it and the fundamental values of these institutions mutually reinforce one another. It is capable of much higher manifestation in more societies and in more domains of life; and we may expect that (though it has opponents—see chapter 11) the trend towards its higher manifestation will continue for quite some time. Its being reflected in the behavior of increasing numbers of people is, therefore, readily explained. But the rational grounds for adopting the modern valuation of control, as distinct from the factors that explain its widespread adoption, rest (I suggest) in large part on the following diverse set of presuppositions. These are often framed by the conviction that techno-scientific innovation will, like it or not, be the leading edge of the trajectory into the future, so that contesting the modern valuation of control is, somehow, out of touch with reality (see section 10.6; section 11.1).[1]

(a) Ongoing techno-scientific innovation expands human potential and serves the well being of human beings in general; and thus it is indispensable for "economic development" and a prerequisite for a just society.

(b) Techno-scientific solutions can be found for virtually all socially significant problems. They are regularly found for major problems, such as disease and meeting energy and communication needs, of significance to virtually everyone; moreover, implementations of such solutions do not generally depend (causally or ethically) on understanding the causal (social) history of the problem (section 8.3). They can also be found for problems occasioned by the "side effects"—for example, health and environmental problems—of techno-scientific implementations themselves (see section 9.2).

(c) The modern valuation of control represents a set of universal values, part of any rationally legitimated value-outlook today, whose further manifestation is de facto desired by virtually all who come into contact with its products.
(d) There are no significant possibilities for value-outlooks which do not contain the modern valuation of control, to be actualized in the foreseeable future.
(e) Natural objects are not per se objects of value, and they only become such in virtue of their places in human practices; per se they can be completely understood in terms of the categories of underlying structure, process, interaction, and law abstracted from any value they may derive from their place in human practices. When we exercise control over objects, informed by sound understanding articulated with these categories, we are dealing with objects as they are in themselves as part of "the material world"—and that is why projects shaped by the modern valuation of control have been so spectacularly successful.

These presuppositions, I am sure, will readily be recognized. For many, they will be regarded as part of the uncontroversial "common sense" of our age.[2] Today, since techno-science is fueled largely by capital and related interests, the presuppositions tend to be linked (often in the business pages of newspapers) with assumptions that support, for example, legitimating controversial technologies and minimizing corporate responsibilities. One such assumption is that, if one makes an investment in a techno-scientific innovation, one is entitled to regain the costs of the investment and to make a profit (though the risk of a loss is never absent). Then, in the process of implementing the technology, risks (to health, environment, etc.) may be sidelined, not by ignoring known risks,[3] but by not accepting the burden to go to great lengths to anticipate theoretically possible risks and to check them out, certainly not to shoulder the burden of costs of evaluating them or of putting burdensome safeguards in place (cf. section 9.4). Then, the call for independent evaluation of risks of controversial novel technologies (which one might have thought is the "scientific" way) is interpreted as a call for research to be funded from public sources; but then, it is added, to call for independent evaluation supported by public funds is to pit public organizations against private innovations, and to create barriers to innovation. The underlying assumption is that granting prima facie legitimacy to the techno-scientific innovations of private enterprise has manifestly brought great benefits to humankind, which far outweigh the harm they may have wrought.

I think that it is quite clear that, if a number of the stated presuppositions cannot be sustained, then the rational grounds for endorsing the modern valuation of control would dissolve—regardless of its widespread embodiment in contemporary social structures and the support it gains from the current hegemony of "globalization." All of the presuppositions, except (e), which may be considered a metaphysical view (section 5.1), are clearly open at least in part to empirical investigation (see chapter 11), which cannot, however, easily be dissociated from value commitments. In the capitalist-oriented world, for example, presupposition (a) is typically supported by individualist views of human nature, which emphasize the individual (agency and the body) and de-emphasize the social character of human beings and their relationships to cultures and groups—human beings as choosers, centers of creative expression, consumers, "preference or utility maximizers," and the like (section 11.2). Then, human well-being tends to be thought of primarily in terms of bodily and psychological health, and the regularly exercised capacity to express a variety of ("authentic" or self-chosen) egoistic values (Lacey and Schwartz 1996). This view of human well-being is contested by, among others, today's anti-globalization movement, and their opposing conception (which emphasizes human agency exercised in the course of participation in communal projects) informs their criticisms of these presuppositions, though not in such a way as to render empirical inquiry irrelevant (section 6.3; appendix: 1, sources, con; chapter 11).

In part II, I will show how these presuppositions, or instances of them that make claims with a narrower reach, play an important role in the current controversy about transgenics, and, in part III, I will argue that it is in the context of this and similar controversies that we can best grasp how empirical inquiry, and what kinds of empirical inquiry, are relevant to their appraisal.

1.2 SCIENCE AS VALUE FREE: IMPARTIALITY, NEUTRALITY, AND AUTONOMY

The view that science is value free is (I submit) well captured by the thesis: Impartiality, neutrality and autonomy are constitutive values of scientific practices and institutions. I will now explicate the three constituents and their respective presuppositions. The succinct accounts presented here should be treated as close approximations; details, nuances, qualifications, variations, and alternative proposals are discussed elsewhere (*SVF*: chapters 4, 10).

Impartiality

Impartiality presupposes that cognitive and other kinds of values can be distinguished (chapter 3). It represents the value to be manifested and embodied in scientific practices that:

> A theory is accepted of a domain of phenomena if and only if it manifests the cognitive values to a suitably high degree, according to the highest standards, in the light of relevant available empirical data; and a theory is rejected if and only if a

theory inconsistent with it has been soundly accepted; and, thus: There is no role for moral and social values (and the uses to which theories may be put, and by whom) in the judgments involved in deciding to accept or reject theories.

Impartiality concerns judgments made when choosing to accept (or not) theories. Note that "accepting a theory" is always an incomplete expression; a theory is accepted or rejected of a domain of phenomena, *and this is to be understood throughout the book, even when it is not made explicit*. Impartiality has no implications about possible roles for ethical and social values at moments when other stances are adopted towards theories, for example, provisional entertainment for the purpose of further testing or deeming them fit for application. To accept a theory (of a domain of phenomena) is to make the judgment that the knowledge expressed in it can be considered settled, not needing further empirical support, so that further inquiry concerning its confirmation or disconfirmation is irrelevant, since all that could be expected from it is repetition of what has already been accomplished. Thus, accepting a theory will back taking it as given in ongoing research and social practice. Sound acceptance, that is, acceptance in accordance with impartiality, while normally accompanied by agreement (within the relevant scientific community) is not constituted by agreement. It requires that—in the light of relevant available data that are representative of data that could be obtained from observing the phenomena (in the relevant domain), and of other actually accepted theories—the theory manifests the cognitive values highly (according to the highest standards) of domains of phenomena whose limits have been thoroughly tested. It requires, in addition, so that it will meet the highest standards (*SVF*: 62–66), that criticism has been exhausted, that is, there are (not that there cannot be, for science does not trade in necessities) no further plausible proposals—after allowing for a suitable lapse of time, and being open to the input (criticisms) and testing of divergent perspectives—of (potential) research projects whose outcomes could put the result into question. The class of settled results is large and rapidly expanding, including those of, for example, molecular chemistry, viral and bacterial causation of disease, electronic theory as applied in such technological devices as computers and television, and classical mechanical accounts of terrestrial motions. But not all results widely accepted among the mainstream scientific community should be considered soundly accepted. This will be illustrated in part two, for example: while it is settled scientific knowledge that maize plants can be genetically engineered so that they produce a toxin fatal to a certain class of insects, the hypothesis that these plants do not pose serious risk to nontargeted insects is not settled, because it has not yet adequately met all the criticisms that have actually been made.

A theory may be soundly accepted (accepted in accordance with impartiality) and at the same time manifest certain social values (e.g., be useful on application for projects shaped by the modern valuation of control); and the same theory may fail to manifest other social values. Impartiality does not forbid a role to social values in making judgments of their "significance"—where a theory is significant for a value-outlook if it may be applied so as to further the manifestation of (some) component values of the outlook, without (on balance) undermining the outlook as a whole (Anderson 2002; 1999; *SVF*: 15–16; see also section 3.2).[4] On the other hand, impartiality does not imply significance (for any value-outlook); and, significance (or expected significance) does not have a role, alongside the cognitive values, in making judgments of whether or not a theory is accepted in accordance with impartiality.

I have represented impartiality as a value of scientific practices and institutions. It is well known, however, that in actual fact, numerous theories have been and are accepted in violation of it; "bias" and "distortion" have marked, and continue to mark, the conduct of science. Nevertheless, their presence is consistent with the adoption of impartiality as a value in scientific practices provided that there are exemplary cases that do manifest it highly (and there are many and I listed a few of them above); provided that the trajectory of scientific practices points to more and a greater variety of theories being accepted in accordance with it; provided that attempts are made, including by way of sociohistorical investigations, to identify the mechanisms that can cause violations of it and steps are taken to prevent their actual operation (section 2.1), ideally steps that are built into the normal activities of the communities of inquiry and the institutions that support them (Longino 1990; 2002). Longino goes so far as to maintain that being the outcome of a certain kind of (social) process of inquiry—involving public recognized venues for appraisal of evidence and methodology, responsiveness to criticism, public standards of appraisal, and tempered equality of participants in research practices (Longino 2002: 129–33)—is constitutive of being a soundly accepted theory. In contrast, I (and Kitcher 2001) maintain only that soundly accepted theories are causal outcomes of such processes (Lacey 2005a). It can be important for the smooth functioning of these processes that a variety of practical maxims be followed: for example, "use double-blind methods in the evaluation of clinical trials," or (in studies with significant social implications) "make sure that a wide range of value-outlooks are represented among the members of the research community."

Neutrality

Neutrality presupposes first that scientific theories do not logically entail that any particular values should be adopted, and second that the body of soundly accepted scientific theories leaves open a range of viable value-outlooks (*SVF*: 74–82; Lacey 1997). Then, neutrality represents the value that:

> Each viable value-outlook is such that (in principle) there are soundly accepted theories that can be significant to some extent for it; and applications of soundly

accepted scientific theories can be made evenhandedly, so that overall (in principle) there are no viable value-outlooks for which the body of theories should have special significance.

Neutrality expresses the value that science does not play moral favorites—that scientific research provides, as it were, a menu of soundly accepted theories, among the items of which (in principle) each value-outlook may have its tastes (good or bad) for application catered to. Proponents recognize that, in actual fact, soundly accepted theories may not be evenhandedly significant across all value-outlooks; the "in principle" is intended to indicate (and this could be added as a third presupposition of neutrality!) that there is nothing in the proper conduct of science that sets it up that soundly accepted theories should be more significant for some rather than other value-outlooks, so that the cause of any actual lack of evenhandedness is to be found in other (social, cultural) factors. It follows from the two stated presuppositions that scientific inquiry cannot adjudicate among currently viable value-outlooks, and thus that science cannot resolve the great moral problems of our times.

Clearly impartiality is necessary for neutrality, but it is not sufficient. Also, it is worth emphasizing, it does not follow from its first presupposition—that an accepted theory has no value judgments among its entailments—that neutrality represents an established fact. Although it is well known that there are theories, developed because of their expected significance for one value-outlook, that also gain significance for another, as illustrated by the fact that many results originally obtained in military research also have valuable applications for peaceful ends (Hacking 1999: 183–84), it remains that whether or not a theory has significance, and for which value-outlooks it has significance, is a matter open to social scientific inquiry. Indeed, as a matter of fact, many items of scientific knowledge are applicable more or less evenhandedly across viable value-outlooks, and many are not.

Like impartiality, neutrality can be maintained as a value of scientific practices despite not always being highly manifested in actual fact, provided that conditions similar to those I listed for impartiality are in place. Within a wide range of mainstream scientific practices the trajectory is indeed in the direction of the higher manifestation of impartiality—and it is easy to point to exemplary cases of theories that are soundly accepted of certain domains. But, in the case of those same practices, a similar trajectory towards higher manifestation of neutrality is not discernible; and, within them, I do not think this can be reversed. Rather than being applied in an evenhanded way, the soundly accepted theories of modern science tend overwhelmingly (and often unabashedly) to be significant for value-outlooks that contain the modern valuation of control, so that actual practices of application provide little evidence that the theories of modern science can be significant to a comparable extent for many other viable value-outlooks. That the mainstream trajectory of modern science is not in the direction of greater manifestations of neutrality is one of the principal themes of this book. It has far-reaching consequences; it underlies my argument for methodological pluralism and is illustrated in the discussion about transgenics and their alternatives in part II.

Autonomy

Autonomy presupposes, first, that there is a reasonably clear distinction between basic and applied scientific research and, second, that the practices of basic research aim to bring about higher and more widespread manifestations of impartiality and neutrality. I represent it as the value that:

> The characteristics of scientific methodology, the strategies adopted in research, and the priorities and direction of basic research are chosen, without "outside" interference, in the light of cognitive interests alone; i.e., they are chosen so that the interest, to heighten the manifestation of the cognitive values in the theories of the domains under investigation and to extend research into domains not currently within their compass, is not subordinated to "outside" interests, powers, and values, be they political, religious, economic, or personal.

Again, no one seriously doubts that actual scientific practices often in actual fact fall short of adequate manifestation of autonomy; and again this is not a decisive reason to reject it as a value of scientific practices. It is always a delicate matter to maintain the distinction between "interference" and "support." Clearly scientific research normally needs the financial (and other) support from institutions (foundations, economic, military, humanitarian, educational) that embody various social, economic, and other values. We may expect that these institutions will condition their support to research on the anticipation of obtaining results that will be significant in the light of the values they embody. So, in order that autonomy be manifested to a reasonable extent, a further presupposition (a complementary value) needs to be added: "Basic research is conducted in autonomous communities, supported by institutions that do not interfere with the methodological decisions and judgments about choice of theories made in the communities; and, in order to prevent that the only projects prioritized in research would have outcomes of significance principally (or only) in the light of specific value-outlooks, overall there is a diversity of ethical and social values represented among the institutions that support scientific research."

Roles for Values Consistent with the Thesis That Science Is Value Free

The thesis that science is value free, I have suggested, may be considered as compounded of the three claims: impartiality, neutrality, and autonomy. So considered, it assumes that scientific practices are permeated through and through with cognitive value judgments, and that they themselves should be held to (though it is not uncommon for them to lapse from) accordance with values, such as the components of science is value free itself, that derive from their aims (section 3.2). Moreover, it leaves open that there are plenty of legitimate interactions between science and ethical and social values, concerning, for example, setting the direction and determining the legitimacy of applied research and applications, the ethics of experimental research, and providing motivations to engage in research on certain problems, even making judgments about the adequacy of available evidence and appraising the testimony of scientific experts (chapter 9; chapter 10). The proponents of "science is value free" readily accept all this (*SVF*: 12–19). They hold only that at its core—where theories are accepted, methodological decisions are made, and research directions of basic research are set—ethical and social values have no proper role. Too often, critics have dismissed "science is value free" by pointing to the kind of play of values in scientific practices that is not questioned by its proponents.

I turn now to a summary appraisal of the three components of "science is value free."

1.3 LINKS BETWEEN ADOPTING STRATEGIES AND HOLDING SOCIAL VALUES—MATERIALIST STRATEGIES AND THE MODERN VALUATION OF CONTROL

Can the ideal, represented in autonomy, that basic scientific research effectively be "driven by" the cognitive values alone, play a regulatory role in scientific practice? It might seem to follow from accepting that the aim of science is to gain understanding of the world; and, operationally, this might appear to amount to the aim "to generate and consolidate theories manifesting the cognitive values highly, progressively of more and more domains of phenomena and possibilities" (section 3.2).

Strategies

When the aim of science is stated like this, however, it can provide no direction to scientific investigation, since it does not point to—for any domain of phenomena—the relevant kinds of empirical data to procure and the appropriate descriptive categories to use for making observational reports, and to the kinds of theories to posit so that they can be put into contact with the data. In order to pursue such an aim the "right" kinds of data and theories must be brought into contact, so much so that (logically) antecedent to engaging in inquiry a strategy must be adopted. Adopting a strategy plays two principal roles. First, it constrains the kinds of theories that may be entertained (e.g., by prescribing the kinds of categories to use in constructing theories), and thus it delimits the domain of phenomena deemed of interest for a research project, and specifies the kinds of possibilities that may be explored in the course of it. Second, reflecting that many of the cognitive values (e.g., empirical adequacy, explanatory power) involve relations between theories and relevant data, it selects the kinds of empirical data that acceptable theories should fit, that is, relevant kinds of empirical data to procure and the appropriate descriptive categories to use for making observational reports—and thus the phenomena and the relevant aspects of them to be observed and experimented upon.

Without adopting a strategy we cannot address coherently and systematically such matters as: what questions to pose in research, what puzzles to resolve, what classes of possibilities to attempt to identify, what categories to deploy both in theories (hypotheses) and in formulating empirical data, what kinds of explanations to explore, what phenomena to observe, measure, and experiment upon, what procedures to use—and also: who are the appropriate participants in research activity, what should their qualifications be, and their life backgrounds and virtues? Unless questions like these are answered, the cognitive values cannot play their role as criteria of choice among competing theories.

What I call "strategy" has much in common with Thomas Kuhn's "paradigm" and "disciplinary matrix," Larry Laudan's "research tradition," Philip Kitcher's "framework" and Ian Hacking's "form of knowledge" (Kuhn 1970—see *SVF*: 261; Laudan 1977; Kitcher 1993; Hacking 1999).[5] Although these notions are not identical, they have arisen in response to common concerns. When any one of them is utilized, it is adopted (logically) prior to engaging in the research it enables and to the appraisal of any theories constructed in the course of that research. Consider Hacking's way of putting it: "By a form of a branch of knowledge I mean a structured set of declarative sentences that stand for possibilities, that is, sentences that can be true or false, together with techniques for finding out which ones are true and which ones false. . . . A form of knowledge represents what is held to be thinkable, to be possible, at some point in time" (Hacking 1999: 170–71). In my formulation a strategy identifies the general features of the possibilities that could be identified by research conducted under it. It limits what is "thinkable," what may be entertained seriously, where it is adopted. Sometimes grasping what is "thinkable" in the context of one strategy will render developments under another strategy unintelligible to those who have adopted the former, that there will be "incommensurability" among competing strategies (4.1). If one strategy gains hegemony in a scientific community or, more broadly, in society, that might lead effectively to limits on what is "thinkable" in that community or society. In such circumstances, adopting that strategy would not be recognized as being in need of rational justification (for the question of why it is adopted rather than some alternative simply is not "thinkable"). Instead it will be considered to be an integral part of scientific methodology and to be built into the aims of science. When a third methodological element, in addition to "theory" and "empirical data," is considered explicitly (whether "strategy" or any one of the others above), questions—both explanatory and justificatory—arise about differences and variations of methodological approach. Hacking, for example, emphasizes how a form of knowledge may be altered by radically new inventions (Hacking 1999: 184). My emphasis is different (though not incompatible with his), that adopting different strategies may be linked with holding different value-outlooks.

Materialist Strategies

Most of modern science tends to adopt virtually exclusively various forms of (what I call) materialist strategies: theories are constrained to those that represent phenomena and encapsulate possibilities in terms that display their lawfulness, thus usually in terms of their being generated or generable from underlying structure (and its components), process, interaction, and the laws (characteristically expressed mathematically) that govern them. Representing phenomena in this way decontextualizes them, by dissociating them from any place they may have in relation to social arrangements, human lives and experience, from any link with value (thus deploying no teleological, intentional, value-laden, or sensory categories), and from whatever social, human, and ecological possibilities that may also be open to them. I will call the kinds of possibilities that can be encapsulated under materialist strategies decontextualized possibilities. A variety of materialist strategies have been deployed over the past four centuries, where the variety arises from additional constraints that might be put on what is to count as a law or on the variables that are admissible in laws.[6] In order that they may be brought to bear evidentially on theories entertained and pursued under the constraints of materialist strategies, empirical data are selected, sought out—subject to meeting the condition of intersubjectivity (and, where appropriate, replicability)—and reported using descriptive categories that are generally quantitative, applicable in virtue of measurement and instrumental and experimental operations. Selecting in this way excludes other kinds of empirical data. To see this, note that any observable phenomenon can be described in indefinitely many ways that meet the condition of intersubjectivity, and many of them involve the use of categories that are not reducible to those deployed under materialist strategies, for example, intentional (as when we describe a person's actions) and some value-laden ones, for example, when we describe an action as skillful or intelligent or a person as suffering acutely (section 3.1). Under materialist strategies, data are reported without using these categories, thus reporting them, stripped from all links with values, in abstraction from any broader context of human practices and experience.

Why has the modern scientific community adopted materialist strategies almost exclusively? Before addressing this, a comment of the phrase "almost exclusively": Sometimes reluctantly, modern science has recognized a place for ecological strategies (using categories like "adaptation"), which have not been reduced to materialist strategies, although the question of whether or not they can be has been opened up in a fresh way following recent developments of mathematical theories of complexity. Also systematic empirical inquiry has been conducted in the human sciences using intentional and value-laden categories. The stereotype prevails, however, that the human sciences, and to a lesser extent ecology, are somewhat deficient in their "scientific" credentials. This helps to explain why research programs, aiming to reduce, for example, behavioral and cognitive phenomena in ways that are amenable to investigation under materialist strategies, tend to be favored in fields like psychology (Lacey 1990; 2003b), and why phenomena—concerning risks, for example, that are relevant to the legitimacy of certain types of applications—are effectively ignored (section 9.5).

Moreover, there are many who take for granted that it is of the nature of "scientific" to be conducted under materialist strategies, that nothing else is "thinkable" within scientific practices. For them it makes little sense to ask why materialist strategies have been adopted almost exclusively. Nevertheless, I am assuming that the possibilities of things are not exhausted by their decontextualized possibilities, and that there are forms of systematic empirical inquiry in which non-decontextualized possibilities can be investigated fruitfully. (I will provide arguments for this assumption in chapter 5 and throughout part II.) For present purposes, note that elsewhere I have identified strategies (of various degrees of development) under which possibilities, which are not reducible to decontextualized ones, may be investigated: for research in agriculture, strategies of agroecology (section 5.4); for research in psychosociobiology, feminist strategies (*SVF*: chapter 9; see also Lacey 2003b); and, prior to the dominance of materialist strategies Aristotelian strategies were dominant (*SVF*: chapter 7). All of this builds on the fact that many objects, including experimental and technological phenomena, whose decontextualized possibilities are well grasped under materialist strategies, are also social objects, objects of potential social value; their possibilities qua objects of social value are not reducible to their possibilities qua objects abstracted from social context (see section 7.2). Understanding them fully (*SVF*: 99–100)—in principle being able to encapsulate all the kinds of possibilities open to them—requires reference to the human/social descriptions that can also be given of their boundary and initial conditions and of their effects, and thus grasping the possibilities that these things have in virtue of their relations with human beings, social conditions, and (more broadly) systems of things (section 5.4; section 9.5; Lacey 2003b). Certain decontextualized possibilities cannot be actualized (in historical context) without also actualizing certain social ones (e.g., furthering the manifestation of the modern valuation of control) and undermining others (section 4.3; section 7.2). To focus on the former and leave aside the latter is to abstract (decontexualize); and to limit the domain of science to the former is just an arbitrary stipulation when systematic empirical inquiry that involves the latter can be conducted.

I will consider any form of systematic empirical inquiry held to accordance with impartiality, whether or not it is conducted under materialist strategies, to be a form of "science" (section 3.2; *SVF*: chapter 5). There are (at least incipient) alternatives, but modern science has been conducted almost exclusively under materialist strategies. Why is this so? (i) What explains it as a socio-historical phenomenon? And, (ii) what are the rational grounds for the almost exclusive adoption of materialist strategies in modern science?

One common answer (to ii) is grounded in materialist metaphysics: Science aims to understand the world as it is—the material world—independently of its relations with human beings; materialist strategies (and only they) provide categories appropriate to this aim (*SVF*: 104–9; Lacey 2002f; see also section 3.3; section 5.3; and Presupposition (e) in section 1.1). A second answer can be drawn from Kuhn's account of the history of science (Kuhn 1970): Not the nature of "the material world," but the current historically contingent stage of our research practices, demands the adoption of materialist strategies. Adopted in the first place (rationally) because they helped to solve puzzles that had remained anomalous under old strategies, they have continued to predominate because of their fruitfulness: under them the range of theories that have become soundly accepted is large and variegated and it continues to become more so. That is sufficient for the current privilege of materialist strategies for, according to Kuhn, the historical practice of science proceeds best when the scientific community pursues a strategy single-mindedly until its potential is exhausted (section 4.3; section 5.2; section 5.3).

Adopting Materialist Strategies and Holding the Modern Valuation of Control

Having criticized these two answers elsewhere,[7] I will proceed directly to a third, the one I take to be most compelling; it involves exploiting links, mutually reinforcing relationships, between adopting materialist strategies and holding the modern valuation of control.[8] My answer is not that materialist strategies are adopted only (or even principally) for the sake of generating applications that further interests cultivated by the modern valuation of control, or because (in the modern scientific tradition) understanding has been subordinated to utility.[9] That would be incompatible with the fact that theories soundly accepted under materialist strategies often provide understanding of domains of phenomena (e.g., in astronomy and paleontology), but there is no genuine possibility of enhancing utility as a consequence of applying the understanding. Adopting materialist strategies for research concerning a specific domain of phenomena may be (and often is) the rational thing to do for reasons that have little to do with control, and research interests may be motivated by values that have nothing to do with exercising control, for example, satisfying curiosity, gaining understanding, cultivating a critical spirit, and other values associated with "basic" or "pure" research.

Even so, adopting materialist strategies and holding the modern valuation of control mutually reinforce each other in the following overlapping ways: (1) Furthering the modern valuation of control is served by and dependent on the expansion of knowledge gained under materialist strategies (techno-science); (2) The contemporary world of lived experience, embodying evermore highly the modern valuation of control, is deeply implicated in interaction with techno-scientific objects, and so this world cannot be understood without the understanding gained under materialist strategies; (3) There are close affinities between technological and experimental control (Dupré 2002: 10–11); and understanding, gained under materialist strategies, is dependent on successfully achieving experimental control; (4) Engaging in research under materialist strategies fosters an interest in the fuller manifestation of the modern valuation of control, since its pursuit often depends on the availability of instruments and equipment that are products of techno-scientific advances; and sometimes technological objects themselves provide models (the clock in classical mechanics, the computer in contemporary cognitive science), or even become central objects, for theoretical inquiry; Finally, (5) in view of current forms of the institutionalization of science—where the institutions that provide material and financial conditions for research are likely to do so because they expect that applications that are "useful" for them will be forthcoming—any values furthered by conducting research under materialist strategies (e.g., satisfying curiosity, etc.) tend to be manifested today within value-outlooks that also include the modern valuation of control.[10]

Recall the question under consideration: Why has modern science been conducted almost exclusively under materialist strategies? (Or: Why is research conducted under materialist strategies widely considered exemplary of—or even constitutive of—scientific research? It is not: Why should a materialist strategy be adopted for research concerning a specified domain of phenomena?) The question really encompasses two: (i) what explains sociohistorically that modern science has been conducted in this way? And, (ii) what are the rational grounds for adopting materialist strategies almost exclusively materialist in modern science?

My answer to (i) is that there exists the mutually reinforcing relations (just described) between adopting materialist strategies and holding the modern valuation of control, and that the modern valuation of control has come to have a grip on leading modern institutions and (more generally) on modern sensibilities. And, if there is a good answer to (ii), it is that these mutually reinforcing relations exist, and that there are good reasons to hold the modern valuation of control—and that research conducted under materialist strategies has been remarkably fruitful (producing a wide range of theories that have been accepted in accordance with impartiality), and it has virtually unlimited potential for further development (section 3.3). Note that fruitfulness and open-endedness are necessary conditions for the adoption (over the long haul) of any strategies, but, when investigating certain phenomena, they do not suffice to justify adopting materialist strategies rather than some others (e.g., agroecological strategies—section 5.4; section 10.1). Elaborating, the modern valuation of control is indispensable, not only for making sense of the social history of modern science, but also for understanding its predominant methodological features. It, combined with widespread adherence to it, explains the virtually exclusive adoption of materialist strategies in modern science and, insofar as there are good reasons to hold it (including that its presuppositions can gain support in empirical inquiry), provides rational grounds for the privilege granted materialist strategies. To answer (i) is not sufficient to answer (ii). The modern valuation of control, in virtue of being widely held and highly manifested and embodied in the prevailing social order, may be the key factor in explaining the predominance of research conducted under materialist strategies; but that, by itself, does not provide a good reason for supporting it. It could only provide a good reason to adopt materialist strategies almost exclusively if there were good reasons for holding it; and those reasons would include that its presuppositions (section 1.1) have gained support from empirical inquiry (section 11.4).

To forestall misunderstanding, I repeat that reasons for the adoption of a version of materialist strategies in a particular area of research need not include holding the modern valuation of control. My thesis is that the virtually exclusive adoption of materialist strategies in modern scientific research is explained by the fact that the modern valuation of control is widely held (and manifested in leading social institutions) and that there are mutually reinforcing relations between holding it and adopting materialist strategies, and that it would be rationally justified if there were good reasons to hold the modern valuation of control. It does not follow from this thesis that all research carried out under materialist strategies is conducted either (narrowly) to serve interests of dominant economic and political groups, or (more generally) for the sake of furthering the modern valuation of control. It also does not follow that the explanation of why a particular version of materialist strategies (e.g., constrained so that laws be deterministic) is adopted, rather than another (e.g., admissibility of probabilistic fundamental laws), must be connected with economic projects, ethical and social values, or any interest in applications. In these cases matters pertaining to fruitfulness may be sufficient. (Metaphysical concerns or disputed ideals of explanation may also be relevant.) The purview of my thesis is the totality of scientific practices and institutions; it concerns the range of strategies that may be adopted in them. Holding the modern valuation of control provides a good reason to restrict research almost exclusively to that conducted under materialist strategies only if there are generally compelling reasons to hold it rather than any competing values. On the other hand, for individual scientists, that they hold the modern valuation of control may provide a good reason for them to adopt materialist strategies in their own research activity. Similarly, that other scientists hold competing values may provide a good reason for them to adopt other strategies.

Values and the Adoption of a Strategy

We may draw out of the preceding discussion the following picture: When one adopts a strategy, in effect one lays out in the most general terms the kinds of phenomena and possibilities chosen to be investigated; in the case of materialist strategies, the decontextualized possibilities of things, and phenomena in spaces where the decontextualized possibilities are the only ones there are. Subject to continuing fruitfulness—constructing and confirming theories that are accepted in accordance with impartiality—being a necessary condition for rationally maintaining one's adoption of a strategy, there is nothing logically improper about social values strongly influencing one's decision about which strategy to adopt. Then, regardless of why the strategy was adopted, the acceptability of theories constructed under it is judged in the light of the data and the cognitive values. The social values in play in deciding to adopt a strategy are not among the criteria of theory choice (although attending to them may help to identify clearly the domains of phenomena of which a theory is soundly accepted).

The important thing is to keep the roles of the social and cognitive values separate (chapter 3). Their different roles reflect different (logical) moments connected with making theory choices. At one moment, when we ask, "What characteristics must theories have to be provisionally considered?" strategies play the key role. Logically they function first. Then (logically) at the second moment, when we ask, "Which (if any) of the theories, that fit the strategy's constraints, is to be accepted?" the play of the cognitive values alone, in the light of the empirical data and other accepted theories, should be decisive. (A third moment, connected with the application of scientific knowledge, is brought into the picture in section 3.4.) Values (ethical/social) have no legitimate role at the second moment acting alongside the cognitive values, and any role that they actually play there is a sign of "bias" or "distortion"; but at the first moment a strategy may be adopted because of mutually reinforcing relations it has with values, with respect to which theories developed under it are expected to have significance. Judgments made at the second moment normally depend upon conditions that derive from judgments made at the first. Nevertheless (logically) the grounds for making them do not rest in any way on the mutually reinforcing relations between strategies and values, although actually being able to recognize the logic in play at the second moment may (psychologically) depend on engaging in research under the strategies. Note also that, as a matter of fact, a theory may not become accepted unless it is (potentially) significant in the light of a particular value-outlook, for it can only become accepted (second moment) if relevant research (prior play at the first moment) is conducted, and—devoid of potential significance for potential funding agencies—the research may not be conducted. Having said this, it remains a serious error to derive the falsity of theories from their lack of potential significance, or from their failure to fit the constraints of favored strategies. Whereas a strategy lays out the general features of the possibilities desired to be encapsulated, a soundly accepted theory encapsulates what the genuine possibilities are (if there are any).[11]

I have maintained the following:

(a) The aim of science cannot be pursued without adopting a strategy.
(b) There is no reason to believe that all the possibilities allowed by phenomena can be encapsulated under a single strategy.
(c) The grounds for adopting (and continuing to adopt) a strategy must appeal to factors distinct from and additional to the cognitive values (whether they be metaphysical, valuative, or simply the subjective interests of the researcher).

And I will illustrate in the next subsection :

(d) Fruitful research being conducted under one strategy may contribute to undermining the availability of conditions needed to conduct research under another potentially fruitful strategy; research cannot be conducted wholeheartedly under all competing (potentially) fruitful strategies at the same time.

It follows from (d) that autonomy is not (even in principle) realizable.[12]

In the picture I am offering, the success of research conducted under materialist strategies contributes to the social consolidation of the modern valuation of control and to its influence in more and more spheres of life. This may largely explain why conducting research virtually exclusively under materialist strategies is relatively little contested, and why it seldom enters mainstream scientific consciousness that there are decisions to be made about which strategies to adopt. Then, scientific research tends to be thought of as the mode of inquiry conducted under materialist strategies—or simply as the next steps in the process of inquiry that has been under way for the past four centuries, the issue of strategies having been resolved with the demise of Aristotelian science (section 5.2); and "being generated under materialist strategies" is treated effectively as another cognitive value (see section 3.2; section 3.3 for why this is a mistake). When science is conceived of in this way autonomy may appear not only to be realizable but also to be manifested to a high degree; and the history of science may appear to unfold in response principally to the cognitive values: Can a theory be generalized? Can it be rendered more parsimonious? Can it be reconciled with another theory? What does it predict under specified boundary conditions? Can its predictions be vindicated experimentally, and can they hold up when we seek greater precision? What new domains of phenomena does it illuminate? What should we expect in new theories in order to address current anomalies? Asking questions like these stimulates a good deal of research, and they are fully intelligible without input from practical or valuative concerns (apart from the fact that the interest and urgency to explain particular phenomena often has practical roots)—provided that there is no contestation of strategies. Even so, it remains that research conducted almost exclusively under materialist strategies cannot manifest neutrality highly, and so autonomy is not rescued, unless it is redefined so that significance is expected not for all value-outlooks but only for those that include the modern valuation of control. While neutrality is often thought of in this way (*SVF*: 236–38), it then becomes neutrality bounded by a value-outlook, and it makes no claim on those who adopt viable value-outlooks that contest the modern valuation of control.

The social consolidation and preeminence of the modern valuation of control does not per se rationally remove interest from competing value-outlooks and from research strategies with which they may be dialectically linked. The picture being offered is consistent with there being a plurality of fruitful strategies, each one in interaction with a particular value-outlook, each one exploring a different class of possibilities, and each one generating theories that become accepted in accordance with impartiality; so that each one enables the reliable encapsulation of possibilities of interest for the respective value-outlooks. Consider agricultural research. Under one kind of strategies, materialist ones, possibilities for agricultural practice opened up by biotechnological research are explored; under another, agroecological strategies, the possibilities of improving organic farming methods by developing "traditional" local agroecological methods. This example will be elaborated in detail, and agroecological strategies will be characterized (section 5.4) and utilized in a variety of ways throughout part II of this book.

Impartiality without Neutrality

The issues raised here are complex. On the one hand, if biotechnological and agroecological strategies were both to turn out to be fruitful it would seem to be an unambiguous gain, enabling us to identify additional classes of possibilities, thus furthering the aim of gaining understanding of "the natural world." On the other hand, it may not be possible to actualize together two genuine possibilities: implementing transgenic technology in agriculture (an application of research conducted under biotechnological strategies) on a significant scale tends to undermine the conditions for agroecology, and vice versa (section 4.1; section 5.6: section 9.5). Why, then, bother to explore possibilities unless they may obtain the conditions to be actualized? Why, given the hold of the modern valuation of control in modern society and on modern consciousness, give attention to agroecology, which runs against the grain of our times (section 10.6)? So, while the picture permits a plurality of strategies, it also helps to explain why there are pressures against actually multiplying strategies. These pressures do not negate the fact that theories accepted under the dominant strategies may be accepted in accord with impartiality, as an ever-increasing number of theories accepted under materialist strategies are. They do, however, put barriers in the path of movement toward greater manifestation of neutrality, and lead to the temptation to treat accord with materialist strategies as effectively just another cognitive value and in violation of impartiality, to reject (or decline to consider seriously) claims made—by agroecologists, for example—simply because they arise under competing strategies (section 5.4–section 5.6; section 10.3).

The issues are even more complex. It does not seem to be coherent even to aim to explore wholeheartedly (let alone attempt to implement simultaneously) all the possibilities of "the natural world." Research requires material and social conditions, and the conditions needed for research under different strategies may be incompatible; even to conduct research exploring one class of possibilities can preclude probing exploration of another class. Research practices can be deeply incompatible, so much so that they cannot be conducted together with integrity in the same social environment (chapter 4). Then, the values highly manifested and embodied in society can be decisive in accounting for what kinds of systematic empirical understanding actually become gained, and thus are available for application (section 10.4; section 10.6).

There is, for example, an imbalance today between the resources devoted to research in biotechnology-informed agriculture and to research in agroecology (developed in detail throughout part II). While this does not challenge the cognitive credentials of the positive discoveries obtained from the former, it makes it unlikely that we will come to gain adequate knowledge pertaining to agroecological alternatives and to the ecological and social "side effects" of implementations of transgenic crops (section 7.2; section 9.5). The biotechnology theories, which have enabled the discovery of the possibilities of producing transgenic crops, may be accepted in accordance with impartiality; and so we may reasonably expect that applying them will be efficacious (section 6.1; section 6.4; section 8.1). The sound acceptance of these theories underlies the efficacy of the applications; and, where the modern valuation of control is highly manifested and embodied, efficacy may be taken practically to suffice for legitimacy (section

9.2; section 10.2).[13] Legitimacy does not derive from efficacy alone, however, but from contribution to human well-being; and so, in the present case, it presupposes the absence both of overriding, undesirable "side effects" and of "better" agricultural alternatives, assumptions that have been challenged by (e.g.) environmentalists and agroecologists (part II, especially chapter 9 and chapter 10). But if one takes legitimacy to be established, there is no motivation to investigate these matters, and it is hard to see how the conditions to investigate them could become available except where that legitimacy is challenged. In this way, the high manifestation and embodiment of the modern valuation of control in the advanced industrial countries helps to explain why gaining understanding pertinent to such efficacy is high on the research agenda, but that pertaining to legitimacy is not—despite the consequent negative impact on the manifestation of neutrality.

Broadening the Horizons of Empirical Inquiry

For many the picture I have sketched is profoundly disquieting. True, it denies ethical and social values a role in making theory choices. But, according to it, the pursuit of basic scientific research does not stand outside of value disputes: the virtually exclusive privilege granted materialist strategies, for example, is only as rational as the adoption of the modern valuation of control. This permits (subject to fruitfulness and, thus, long-term empirical constraint) the adoption of alternative strategies in view of their having mutually reinforcing interactions with value-outlooks that contest the modern valuation of control—apparently opening choice of strategy potentially to contestation like that currently found about matters of values. Be that as it may, granting privilege to materialist strategies currently lacks rational grounding. My picture pushes us to broaden the horizons for empirical inquiry.

How does it do this? Well, recall the question: Why grant privilege to materialist strategies? The answer is because they are fruitful and they interact in mutually reinforcing ways with the modern valuation of control. Why hold the modern valuation of control? Part of the answer involves appeal to its presuppositions. But, for the most part, they cannot be investigated empirically under materialist strategies (since they deal with nondecontextualized possibilities), and so research conducted exclusively under these strategies cannot feed back and provide them with support or criticism (chapter 11; Lacey 2003b). We need a plurality of strategies for that.

Earlier I defined impartiality, a value of scientific practices: accept theories if and only if they manifest the cognitive values to a high degree. Commitment to neutrality involves extending the range of domains of which we are able to accept theories in accordance with impartiality—towards the ideal, I suggest, that any belief that plays a role in informing or legitimating any salient contemporary social practice be subject to appropriate empirical scrutiny. When research is conducted virtually exclusively under materialist strategies, we gain numerous soundly accepted theories that inform numerous efficacious applications (generally of interest where the modern valuation of control is held), but rarely does it speak directly to the presuppositions that legitimate the modern valuation of control or, for example, to agricultural practices that are expressions of it (part II). Must we then conclude that covertly these presuppositions are accepted because they serve to legitimate the modern valuation of control—that social values are in play alongside or in place of the cognitive values, and thus that the presuppositions are not accepted in accordance with impartiality, and thus that they are ideological?

Ironically, or paradoxically, one could perhaps cut through this impasse by permitting research under a plurality of strategies. But the conditions required to carry out such research may not be readily available where the modern valuation of control is highly manifested, so that to carry it out may already involve commitment to a value-outlook that contests the modern valuation of control. Nevertheless, unless research is conducted under a plurality of strategies, any value-outlook must rest upon key presuppositions that have not been accepted in accordance with impartiality. Then, neutrality will not be an approachable ideal.

Endorsing Neutrality?

The pull of neutrality includes the desire to minimize the likelihood that one set of values should dominate the direction of scientific research. The mainstream tradition has thought that the way to achieve this is to conduct research under materialist strategies, which permit no value-laden terms to be deployed in theories. Doing this certainly ensures that scientific theories have no value judgments among their entailments (as required by one of the presuppositions of neutrality), but it does not ensure evenhandedness of significance across value-outlooks on application. My alternative is to multiply strategies so that the interests of numerous value-outlooks can come to be informed by the sound results of scientific research, and so that key presuppositions of value-outlooks that influence the adoption of particular strategies can be investigated empirically. This is my version of methodological pluralism.[14] It raises important questions about who are members of the communities engaged in systematic empirical (scientific) inquiry, about whether they include only "professionals" with specified qualifications, and about whether it is also appropriate to require that there be a spread among the members of the communities of value-outlooks that are held (Longino 1990; 2002; *SVF*: chapters 4 and 10, discussions of autonomy). It also leads to posing questions about the place of science in democracy (Anderson 2002; section 9.5; section 10.3).

Consider, for example, Presupposition (d) of the modern valuation of control (section 1.1): "There is at the present time no genuine possibility for the high social embodiment of value-outlooks that do not contain the modern valuation of control" (see chapter 11 for how to investigate a proposition of this kind). It is backed by various more specific assumptions, for example, one deployed to legitimate the accelerated implementation of biotechnology-informed agriculture: "There are no significant lost possibilities occasioned by that implementation; apart from it not enough food will be produced to feed the world's expanding population" (a version of proposition P_4, that is used in part II in arguments for the legitimations of using transgenics; see chapter 10). Now, research conducted under materialist strategies can produce answers to questions like: "How can we maximize food production under 'optimal' material conditions?" But not to questions like: "How can we produce food so that all the people in a given region will gain access to a well-balanced diet and so that social and ecological stability will be sustained?" (*SVF*: chapter 8). But "no lost possibilities" can gain no empirical support apart from attempts to address empirically questions like the second one. Thus unless strategies (of agroecology) are defined under which these questions become addressed in a systematic empirical way, assumptions like "no lost possibilities" must remain essentially ideological in character (section 10.4). Strategies of agroecology, which incorporate materialist strategies in a subordinate role, aim to grasp farming practices without dissociating from the social and ecological relations into which they enter—and they sometimes exhibit continuity with the "traditional knowledge" of a culture (section 5.4). They might not turn out to be fruitful, but we cannot know that in advance of engaging in research under them (section 10.1).

Let me reiterate that my argument is against granting exclusivity to research conducted under materialist strategies. At the same time, I do not consider materialist strategies simply to be on a par with other strategies. Their extraordinary fruitfulness precludes treating them in this way. More importantly, control of natural objects plays a role in all forms of life regardless of value-outlook adopted—the modern valuation of control is not adopted in all value-outlooks but, where it is not, control subordinated to other social values is, so that we reasonably expect that, for any viable value-outlook, some products of research conducted under materialist strategies will be significant to some extent. Thus, research conducted under materialist strategies, appropriately subordinated to the desired social values, and not adopted, for example, to the exclusion of agroecological strategies, will probably remain an indispensable constituent of research under all strategies.

The definition and development of a plurality of strategies, including agroecological ones, would further neutrality (section 10.4; section 11.4). Some of the strategies, as we have seen, involve mutually reinforcing interactions with various value-outlooks that contest the modern valuation of control—those of some environmental groups and those who hold the values of "popular participation" (section 6.3). If these strategies prove to be fruitful, their products would contribute to furthering the manifestation of these values—providing further grounds for challenging the modern valuation of control. Neutrality and the modern valuation of control may pull in opposite directions.

Summary

Is science value free? It will now be apparent that a "yes or no" answer would be misleading. Rather we need to discern the ways in which science is and is not value free. Here is my conclusion: Autonomy is not realizable. Neutrality is susceptible to fuller manifestation in the practices of science (systematic empirical inquiry), but only if a plurality of strategies are adopted and research under them actively pursued. The mainstream trajectories of science, constrained as they are for the most part to adopting materialist strategies, therefore, do not promise to bring about a fuller manifestation of neutrality, so much so that one may query whether modern scientific practices today remain committed to the furtherance of neutrality. But impartiality remains a key value of all research practices conducted under any strategy.

NOTES

1. While revising this passage, I encountered the following remarks in editorial-page articles in major newspapers: "Americans are smitten by the idea that new technologies will revolutionize life as we know it and greatly expand human potential" (Staples 2004). "[People in modern Western cultures] are convinced that the future can be shaped by the thrusts of technology. . . . [And] the majority of conventional scientists are prisoners of a notion of time in which the future is always under the control of human beings" (Leite 2003, my translation). Staples goes on to point out that, frequently, valued new potentials quickly become overwhelmed by uses that are not valued (his example: communications breakthroughs leading to the deluge of "spam"); and Leite points to the reflection of a Xavante chief who is seeking to balance the insights of modern science with his culture's traditional view that much of nature must remain outside of human control.

2. This "common sense" makes virtually irresistible (what I call) "the techno-scientific mindset." This is the mindset that tends to follow the following steps: (1) When there is a problem at hand—concerning health, food production, energy, communications, or whatever—look to techno-science and its innovations for the solution. (2) Such innovations and solutions have prima facie legitimacy, which may be questioned, but then the burden of proof falls on the opponent. (3) Risks are a secondary consideration, to be dealt with (except when subject to legitimate regulations) when harms arise. (4) If "unintended" harms arise, we can expect to be able to deal with them adequately and promptly by further developments of techno-science. (5) When a proposed techno-scientific innovation is considered beneficial to its proponents, there is no need to consider alternatives, certainly not alternatives that are not themselves techno-scientific products. (6) Techno-scientific solutions to a problem generally do not draw upon knowledge of the social causal history of the problem (e.g., techno-scientific proposals to deal with vitamin shortage in the diets of poor children—genetically engineer sources of the vita-

min into rice they eat—have been made, which bypass completely investigation of how the problem arose, section 8.3); and implementing or looking for such solutions (e.g., in fetal stem-cell research) should not be curtailed by "intrusive" governmental regulations or by ethical reservations that derive from value-outlooks (e.g., certain religiously based ones) that have been "surpassed" by the novel potentials introduced by techno-scientific developments.

3. The discovery (December 2003) of a case of "mad cow" disease in the United States has raised many issues connected with this point. Several newspaper articles (a good example is Milloy 2004) have opposed additional regulation of the beef industry, claiming that there is no evidence available of transmission of the disease from cows to humans and that the most widely accepted "hypothetical" mechanism (prions) has not been well established. In the absence of demonstrated risk, no regulatory action is necessary (cf. discussion of the "precautionary principle"—section 9.4).

4. Kitcher (2001) introduces a different (but related) notion of "significance," one with cognitive as well as social components.

5. This is related to my engagement with Kuhnian themes in chapters 3 and 4 (and also in *SVF*: chapter 7).

6. For more detail, see the discussion of materialist strategies in section 3.3; also Lacey (2003b). I see no difficulty in re-formulating my account to fit with those views (e.g., Giere 1999: chapter 5) that consider theories to be abstract models (or sets of models) that represent aspects of the world (phenomena and the possibilities they admit), where the components of the models have quantitative properties and are structured in such a way that their processes and interactions exemplify "mathematical principles." Then, mathematical principle, rather than law, would be the core notion of materialist strategies.

7. There is at least one other proffered answer (a neo-Cartesian one) that I will only mention here: Materialist strategies express an ideal of intelligibility. Results obtained under them are "clear and distinct"; they have a kind of definiteness or precision (residue of the old cognitive value of certainty) that is unrivalled in other spheres of social life.

8. This argument is developed in detail in *SVF* chapter 6; it is restated, consolidated, and brought to bear in my analysis of the transgenics controversy in various ways throughout this book.

9. Bacon, perhaps, wanted to subordinate understanding to utility, and what I am calling "the modern valuation of control" has its historical roots in Bacon's deliberations on "utility." My position may be considered neo-Baconian.

10. The view being articulated here is not new. It has resonance, for example, with themes in an article by Thorstein Veblen, originally published in 1906, which I will quote at length:

> In so far as touches the aim and the animus of scientific inquiry, as seen from the point of view of the scientist, it is a wholly fortuitous and insubstantial coincidence that much of knowledge gained under machine-made canons of research can be turned to practical account. Much of this knowledge is useful, or may be made so, by applying it to the control of the processes in which natural forces are engaged.... The reason why scientific theories can be turned to account for these practical ends is not that these ends are included in the scope of scientific inquiry. These useful purposes lie outside of the scientist's interest. It is not that he aims, or can aim, at technological improvements. *His inquiry is as 'idle' as that of the Pueblo myth-maker. But the canons of validity under whose guidance he works are those imposed by the modern technology, through habituation to its requirements; and therefore his results are available for the technological purpose. His canons of validity are made for him by the cultural situation; they are habits of thought imposed on him by the scheme of life current in the community where he lives; and under machine conditions this scheme of life is largely machine-made.* In the modern culture, industry, industrial processes, and industrial products

have progressively gained upon humanity, until these creations of man's ingenuity have latterly come to take the dominant place in the cultural scheme; and it is not too much to say that they have become the chief force in shaping men's daily life, and therefore the chief factor in shaping men's habits of thought. Hence men have learned to think in the terms in which the technological processes act. This is particularly true of those men who by virtue of a particularly strong susceptibility in this direction have become addicted to that habit of matter-of-fact inquiry [i.e., inquiry conducted under materialist strategies] that constitutes scientific research. (Veblen 1919: 16–17, my italics)

11. The last two paragraphs quote from and paraphrase *SVF*: 231–34.

12. See *SVF* chapter 10 for discussion of a circumscribed version of autonomy.

13. The fourth component of the modern valuation of control (section 1.1) may be interpreted as having this implication.

14. Other versions of methodological pluralism have recently been proposed (Kitcher 2001; Longino 2002). My version is more encompassing than Kitcher's (which does not take into account explicitly the possibility of conducting research using strategies that are not reducible to materialist ones); and, unlike Longino's, it does not run the risk of compromising impartiality that comes from her considering (as mentioned in section 1.1) that being the outcome of a certain kind of social process is constitutive of soundly accepted theories (Lacey 2005a).

Chapter Two

Objectivity and Serving Human Well-Being

Linked with the idea that science is value free is the idea that science aspires for, and characteristically achieves, objectivity. Can an ideal of objectivity be sustained after the analysis of the interplay of science and values proposed in chapter 1? How can one aspire for objectivity and, at the same time, opt to engage in research whose strategies are adopted in view of their mutually reinforcing relations with one's value-outlook (and its associated ethical ideal)? Does not the methodological pluralism that I have urged undermine any aspiration to objectivity? In this chapter, I propose not only that methodological pluralism enables a robust sense of objectivity to be sustained, but also that the ideal of objectivity actually points toward it (Longino 1990) rather than toward conducting scientific research virtually exclusively under materialist strategies.

2.1 OBJECTIVITY

I suggest that objectivity be understood as the outcome of commitment to impartiality and neutrality. Hence, achieving it requires avoiding the pitfalls that can threaten impartiality and adopting the appropriate plurality of strategies (or methodological approaches) necessary for bringing neutrality to fuller manifestation in scientific practices. Doing these things requires keeping the moments of strategy adoption and theory choice clearly distinct and defining the strategies under which research is being conducted with sufficient explicitness so that the reasons for their adoption can be readily formulated and appraised.

Mechanisms Underlying Departures from Impartiality

Striving to further objectivity requires that attention be paid to empirical sociohistorical studies of the mechanisms underlying departures from impartiality and testing that one's judgments of theory acceptance are unaffected by these mechanisms. These mechanisms have their origin in the social character of scientific inquiry. Any judgment of accepting (or rejecting) a theory, including those that accord with impartiality, is also the outcome of a social process, an outcome of interactions among investigators located in various institutions that (ideally) embody values considered appropriate in view of the aim of scientific practices. Such social processes may lend themselves to embodying additional other social values. Judgments that accord with impartiality can be made only when the process of inquiry is so constituted that the significant causal factors leading to judgments of acceptance of theories do not contain any of the mechanisms that underlie departure from impartiality. As mentioned in section 1.2, not all accepted scientific results do accord with impartiality. Social and historical studies of science can help us discern which ones do and which do not, and provide explanations of why certain theories are accepted by members of the scientific community despite violating impartiality. In this way the empirical scrutiny of scientific practices and the making of scientific judgments can contribute to the sound cognitive appraisal of theories. (Even when a theory is accepted in accord with impartiality, there remains much to be explained by social studies—*SVF*: 231–36.)

There are, of course, cases of outright dishonesty and fraud, of threat and manipulation by the powerful, and of bias that overshadows inquiry and dulls the critical powers of an investigator. Normally, however, we would expect the mechanisms of departure from impartiality to be subtle and difficult to detect, for no one who purports to be doing science will willingly (or candidly) violate impartiality. Anticipating the discussion of part II, it will be helpful to identify some of the more subtle mechanisms, often acting in concert, that are in play in the controversies about transgenics. One is particularly worthy of note, where a social value subtly (usually without awareness) comes to assume a role alongside (or above) the cognitive values when judgments are made about accepting and rejecting theories or hypotheses. This can happen because (as we have seen) theories and established scientific knowledge may be evaluated both for their cognitive value and for their significance or the potential social value of their applications; and because—since, in the daily activities of the researcher, conducting research may be deeply intertwined with efforts to acquire the material and social conditions needed for the research—it may not be easy to separate cognitive from some other values in their actual manifestation, for under these conditions judgments of acceptance and significance effectively will tend to coincide.

The following statement provides an example: "By its very nature, biotechnology tends towards privatization" (Postlewait, Parker, and Zilberman 1993). In making this statement the authors purport to be simply stating a fact (apparently for them with no methodological implications) about the current situation in which scientific research is increasingly becoming intertwined with corporate interests. Given the material and social conditions under which biotechnology research currently is proceeding, acceptance of biotechnology theories (in accord with impartiality) may be virtually inseparable from appraising them as especially significant for projects that express values linked with privatization. As such, this is compatible with impartiality. But it is easy to slip from this to rejecting theories (e.g., in agroecology—section 5.4; section 10.4) on the ground that they do not serve the interests of privatization, or (relatedly) to consider criticisms of privatization to be objections to accepting biotechnology theories that are in fact soundly accepted; and this is not compatible with impartiality. From the fact that a soundly accepted theory serves the interests of privatization, it does not follow that a very different theory (developed under a competing strategy) and serving competing interests, is not also soundly accepted, even if the possibilities it is able to identify cast doubt on the legitimacy of some practices of biotechnologically-informed agriculture. Where the manifestations of cognitive and certain social values appear to be virtually inseparable, this may be too fine a point to be genuinely sensitive to it. That is how the mechanism under discussion operates (section 5.4; section 6.3; section 10.5). It follows that objectivity would be well served by following the maxims: "Make every effort to separate judgments of acceptance of theories from judgments of their significance for particular value-outlooks;" and "When a soundly accepted theory is significant for one's value-outlook, scrutinize rigorously the rejection of any theories that (if soundly accepted) would be significant for rival value-outlooks or that would tend to undermine the legitimation of practices informed by one's value-outlook."

Another mechanism that can underlie departure from impartiality is, de facto, to treat adopting a strategy as one of the cognitive values. Sometimes the play of this mechanism is quite explicit (and justifications for its use are offered), especially in some well-known cases in which consistency with a religious or metaphysical viewpoint is a feature of a strategy's constraints: consistency with specified Biblical interpretations, in the Church/Galileo episode; with dialectical materialism, in the Lysenko affair; and with materialist metaphysics, in the common insistence in cognitive science that intentional explanation must be reducible to or replaceable by accounts that are derived under versions of materialist strategies. Consistency with a metaphysical viewpoint (of any variety) is not a cognitive value (section 3.3; Mariconda and Lacey 2001), though it may serve as a heuristic connected with stances that may be taken toward theories prior to their acceptance or rejection. When a metaphysical viewpoint has a pervasive grip on the imaginations of a culture, it may be difficult to discern this mechanism in play; and when it is endorsed by a repressive power it may be difficult to resist. It is even more difficult to discern when it is virtually uncontested in the mainstream of science that research conducted under certain kinds of structures (materialist strategies) is exemplary. Then it becomes easy to dismiss results gained under other strategies with barely a thought (section 5.4;

part II). A symptom of this mechanism being in play is that results are over-generalized, and a subset of possibilities (decontextualized ones) is identified with the totality of possibilities.

Yet another mechanism—one we will encounter when discussing hypotheses pertinent to assessing the legitimacy of transgenics (chapter 9; chapter 10)—is to infer from the absence of evidence against a theory (hypothesis) that it is acceptable in accordance with impartiality (or, at least, that there is a strong presumption in its favor). Such inferences are unsound unless there has been adequate consideration of questions, such as: Has a sufficient variety of kinds of data been gathered? And has a sufficient range of theoretical competitors been worked out fully enough to be put into serious competition with it? Has the degree of manifestation of the cognitive values in the theory, especially empirical adequacy, been established against the toughest standards (*SVF*: 62–66; section 9.3–section 9.6)? Relations of power may account for the absence of well-worked-out alternatives, rather than the cognitive deficits of nascent alternatives; or from the perspective of alternative value-outlooks and the strategies linked with them, certain kinds of "data" may be simply uninteresting. Where research is conducted virtually exclusively under particular strategies that are dialectically linked with particular social values (as I maintain materialist strategies are with the modern valuation of control), answers to the questions just posed are likely to be given prematurely, especially when the answers are "convenient" for interests linked with the social values (see chapter 9; chapter 10).

Objectivity and Pluralism

Furthering the achievement of objectivity also requires that the community of investigators permit—and nurture by seeking to ensure that a variety of value-outlooks are well represented among them—a plurality of strategies to be developed in research practices in fields where it is appropriate. The community should do this with full awareness of how adopting a strategy may be dialectically linked with holding a particular value-outlook (as well as being subject to the test of long-term fruitfulness) and of the difficulties of balancing and prioritizing the claims for support of research conducted under competing strategies. Where this kind of pluralism is institutionalized, we might anticipate (see section 10.4 for what might be implied in a concrete instance): (a) Values will not play a covert role in accepting and rejecting theories; (b) Value disputes will not be dissociated from the discourse of the worldwide body of scientific investigators—so that the presuppositions of competing value-outlooks can become objects of scientific inquiry (rather than matters of ideological convenience); (c) Scientists will become free to adopt strategies because of their dialectical links with interests and values of, for example, impoverished peoples and their emancipatory movements; and, with provision of appropriate material and social conditions, to put the question of human well-being at the center of the scientific agenda, by asking about the likely (and potential) impact on human welfare of application of scientific knowledge, regardless of whether it is gained under their favored strategies or others; (d) Science will be institutionalized to enable research to be conducted under a sufficient range of strategies, so that there is movement towards each currently held viable value-outlook being able to draw (more or less evenhandedly), from the body of soundly accepted theories, those that are significant for their respective interests; Finally, (e) scientific institutions will be open to the input of democratic deliberations.

In addition, furthering objectivity will balance (and strengthen) the well-known notion that objectivity involves cultivating among scientific practitioners the virtues of "scientific ethos"—honesty, disinterestedness, forthrightness in recognizing the achievements of others and opening one's own contribution to their critical scrutiny, and courage to follow the evidence where it leads (Merton 1957; Cupani 1998; Mariconda and Lacey 2001)—by pointing to the need to cultivate also the virtues of tolerance and openness to knowledge gained under strategies dialectically linked to a multiplicity of moral and social values. Then, it will be difficult for expertise to function as a cover for a shared value-outlook.

Keeping Us Honest and Serving Humanity

When we hold that objectivity is the outcome of commitment to impartiality and strategic (methodological) pluralism, we see that there is no necessary incompatibility between the quest for objectivity and conducting scientific inquiry for the sake of furthering human well-being (Lacey 2005b). Avoiding the pitfalls that threaten impartiality and cultivating strategic pluralism would both keep us honest and serve humanity. So long as we insist on impartiality of scientific judgment, the interpenetration of science and ethical/social values should not undermine that, in scientific inquiry, we gain reliable knowledge of the causes and possibilities of phenomena of the world.

Objectivity, I have argued, points toward research being conducted under a plurality of strategies. Commitment to objectivity, however, may generate tensions between the interests of science as a worldwide community of investigators and the value-outlooks of individual investigators. These tensions arise from the mutually reinforcing relations that often exist between adopting a certain kind of strategy and holding a particular value-outlook. Then, investigators will opt quite reasonably to conduct research under strategies that are linked with the value-outlooks that they themselves hold, that is, to engage in significant research, research that produces results that not only accord with impartiality and contribute to solve puzzles that have arisen under their adopted strategies, but also have social value. Given the competition for the scarce resources that are available for supporting scientific research, however, strategies linked with value-outlooks which are held by minority viewpoints or interests, are likely to remain underdeveloped and considered marginal, all the more so when a hegemonic value-outlook is held that dims awareness that there are both value and strategy options. In this way, the cognitive interest of scientific practices in strategic pluralism may be pushed aside so that the predominant strategy may monopolize.

Professionalism

Another factor also comes into play here: professionalism. Professional interests (for positions, careers, research support, awards, promotions, income, prestige) tend to incline one to engage in practices that are well within the scientific mainstream, and scientific education tends to form scientists with this in view. Scientists are formed to engage in what scientists are actually doing, and what mainstream scientists actually do is to engage in research conducted almost exclusively under materialist strategies (where, to their own minds, only choice of the specific problems for research is affected by the values of the institutions which support or employ them). In discussions, I have often heard the following response to my arguments: That is what scientists actually do; and what scientists do is what science is. Scientists, so far as they are aware, are just carrying on the professional activities for which they have been trained in reputable educational institutions and whose legitimacy "our" society (which functions democratically) has clearly sanctioned. They are simply following through on the latest options provided under materialist strategies without questions about strategic options being posed. If there is no choice of strategy, then criticisms of engaging in research conducted virtually exclusively under materialist strategies, and pointing to the cognitive virtues of pluralism, will appear to be criticisms of engaging in scientific research per se,[1] and as having the effect of threatening funds for research (or even threatening the livelihood of scientists). After all, the vast majority of professional scientists have not been trained to take pluralism, and the issues about values that accompany it, into account; and often they see proponents of pluralism (and other critics of particular research projects being conducted under materialist strategies) as challenging what they do, but without having the means to provide them with alternative ways to engage in their professional activities. (Agribusiness corporations, in ways that cannot be matched by agroecological projects, offer employment opportunities to large numbers of scientists trained in biotechnology.) My arguments, it is said, fly in the face of what scientists actually do and what they are trained to do. Whether this response is made smugly or with anguish (I have encountered both), it does point to a genuine personal predicament in which scientists might find themselves caught and, as such, it should not be dismissed cavalierly or with hostility. Professionalism does contribute to explain that materialist strategies are both predominant and largely unchallenged as such. But it does not provide good reasons, grounded in the character of scientific practices, for why this is so.

Scientific practices may be looked at in two ways: on the one hand, they are committed to the furtherance of certain values that derive from the aim to gain understanding of things and phenomena; on the other hand, they are actually enacted in specific ways that reflect professional interests and societal support. The response to my arguments mentioned above identifies science with what scientists actually do; my arguments identify it with its driving values. There may be (and I believe that there actually is) a gap between the two.[2] It may be reduced either by subordinating the cognitive value of science to professional interests, or by furthering the institutionalizing of pluralism so that cognitive value is held in foremost consideration. Opting for the former way lessens the likelihood that barriers will be put in place to counter the mechanisms of departure from impartiality—perhaps not across the board, but when dealing with hypotheses (like those needed to legitimate applications) where the outcomes of research could challenge interests that are allied with the scientists' professional ones. My arguments spring from taking the latter way. They may count for little, where the professional interests are entrenched, and so may actually have little impact on the conduct of science. Be that as it may! But then, neutrality and the objectivity I have defined should not be articulated as values of scientific practices. The consequence is that science is, then, unable to address adequately such issues, pertaining to the legitimacy of applications, as there being no serious risks and no better ways.

Let me digress and note in passing, though I will not develop the point, that at the present time there are striking new trends in contemporary scientific practices, and the institutions in which they are carried out, that may lead to profound rethinking of the nature of science, especially if one insists that science is what professional scientists do. They may even lead to the ideas of science as value free and objective becoming seen as relics of a past era. These trends are especially apparent in the biological sciences (e.g., concerning the human genome project as well as developments of transgenics) where the components of the underlying order (e.g., genes) are increasingly becoming thought of as objects to which one may hope to obtain intellectual property rights; thus becoming thought of, among other things, as historically constituted property and commodities. With this, the institutional intertwining of scientific research and the projects of corporations is being ever more fostered and made into a reality. This is likely to have far-reaching effects on conceptions of science and scientific methodology. It is hard, for example, to see how neutrality will maintain its appeal as an ideal; and one must wonder what will become of impartiality—and, ironically, even the central place accorded to materialist strategies—when research becomes hindered by the exercise of ownership rights over, for example, gene sequences, and research results become components of advertising and marketing. The implications of this new phenomenon of scientific practices need to be explored more fully and carefully. (End of digression.)

The social values that are held may have profound impact on what strategies become developed and gain the opportunity to display (or fail to display) their fruitfulness. But the cognitive interests of science should incline one toward strategic pluralism. With strategic pluralism, a fuller range of possibilities open to phenomena can be investigated. This makes possible movement toward the higher manifestation of neutrality, and—when the range includes possibilities that arise in connection with discussions of risks of, and alternatives to, applications—hypotheses crucial to the legitimacy of applications can be given the kind of investigation that is desirable (chapter 9; chapter 10). Strategic pluralism probably can only be sustained if a plurality of value-outlooks is represented within the scientific community. Again, given scarce resources, implementing pluralism will most likely diminish the intensity of research conducted under the predominant strategy. The tension cannot easily be avoided, and it is reflected in the version of the aim of science that I offer in section 3.2.

The Unity of the Natural and Social Sciences

The cognitive interests of science support strategic pluralism in another way, too. Social values are objects of contestation. But I have maintained (section 1.1) that holding a value-outlook has an array of presuppositions on which its rational credibility depends. It serves the cognitive interests of science that these presuppositions be subjected to empirical investigation; according to my version of the aim of science (section 3.2), no phenomena of significance in human experience or practical social life are (in principle) excluded from the compass of scientific inquiry. This is especially true concerning the presuppositions of the value-outlooks that are dialectically linked with the adoption of research strategies. But, the presuppositions will be excluded from scientific inquiry unless research is conducted under an appropriate plurality of strategies.

Consider materialist strategies. Adopting them and holding the modern valuation of control mutually reinforce each other. Rationally holding the modern valuation of control (as distinct from subjectively preferring its component values) cannot be upheld apart from belief in a set of presuppositions (listed in section 1.1) that clearly cannot be investigated under materialist strategies, for they do not (and cannot) dissociate the phenomena they deal with from human experience and practical social life. Whether or not theories accepted under materialist strategies accord with impartiality has nothing to do with whether or not the modern valuation of control is held rationally or with whether or not its presuppositions can be sustained by empirical inquiry. But the reasonableness of considering research conducted under materialist strategies exemplary and to be emulated (to the extent possible) in all fields of scientific inquiry does depend on these things, as does considering other forms of systematic empirical inquiry to be somehow "not quite scientific."

In principle, research conducted in the social sciences could disconfirm these presuppositions (see chpter 11) and thus cast doubt on the significance of many results obtained under materialist strategies—though not all of them, for I have maintained that materialist strategies do have a (at least a subordinate) role in connection with all strategies. Of course, the research might confirm the presuppositions. Social science research thus potentially has impact on what research is significant in the natural sciences, on what research is worth doing (in the light of limited resources). At the same time, continued fruitfulness of research conducted under materialist strategies, and continued success in embodying the modern valuation of control in the leading institutions of society (nourished by the products of that fruitful research), will feed back and provide data that is likely to provide some empirical support for the presuppositions. But further embodiment of the modern valuation of control alone is insufficient to confirm all the presuppositions, for example, Presupposition (d), that there are no possibilities for value-outlooks, which do not contain the modern valuation of control, to be actualized in the foreseeable future (section 11.1).

The point here is readily generalized: adopting a strategy is often dialectically linked with holding a particular value-outlook, whose rational credibility depends (at least in important part) on presuppositions that are open to confirmation or disconfirmation in the social sciences. (I show how this applies to agroecological strategies, the values of popular participation, with which they are linked, and the presuppositions of holding these values in section 11.) Then, the significance of results obtained under the strategy is dialectically linked with results gained in research conducted under appropriate strategies in the social sciences, while successful application of results gained under the strategy is likely to constitute (partial) evidence for some of the presuppositions of the related value-outlook. When this is recognized, a profound unity between the natural and social sciences, based on the dialectical relations between them, appears to be possible (Lacey 1990; 2005b). Contrary to a common viewpoint, reducibility and disunity cease to be the only options concerning relations between the natural and the human sciences. The appeal of reducibility remains strong where the modern valuation of control is highly ranked in the dominant value-outlooks of a society (or of an epoch); for then, instead of developing the strategies needed to investigate the presuppositions of holding the modern valuation of control, the push to extend the predominance of research conducted under materialist strategies (deploying new versions of them) into the human sciences is great. On the other hand, where the modern valuation of control is challenged, inquiry in the human sciences will contribute to determining the significance of theories confirmed under materialist strategies, to directing the focus of research conducted under them, and it may lead to interpreting many of the theories so confirmed as providing understanding of limited spaces and domains of objects, mainly techno-scientific ones, of interest only where the modern valuation of control is held.

The dialectical relations between the natural and social sciences are often obscured. They will be obscured wherever the adoption of materialist strategies (and only them) is seen to be virtually essential to scientific inquiry. The notorious difficulty of obtaining agreement in the social sciences, and widespread resignation that impartiality is practically not realizable in them, also further the obscuring. Nevertheless, that these relations are obscured does not undermine the argument that they exist. A symptom of their existence, even when obscured, is that where materialist strategies are adopted without explicit argument for the significance of anticipated results, and these results are applied as a matter of course, we find that the presuppositions of the modern valuation of control (or some of them) tend to be among the reigning convictions of the institutions that implement the applications, as if they have been established beyond reasonable doubt; in these circumstances they function as part of the "untested consciousness" of a culture (section 9.2; section 10.2; section 11.4). The price of the grip of untested consciousness is that matters of significance become subordinated to matters of efficacy, with consequent limitations imposed on objectivity. The solution is to permit an appropriate plurality of strategies to be adopted within the community of investigators, but now understood as including recognition of the dialectical unity of the natural and the social sciences, and restructuring scientific practices to express this unity.

2.2 SCIENCE AND HUMAN WELL-BEING

Questioning the claim that science is value free is often interpreted, from the mainstream of modern science, as expressing an "anti-science" stance (see discussion of "professionalism" in section 2.1). And no doubt some, who question it, do desire to deny any special objective status to scientific knowledge, to undermine its authority, and to portray it as just another object of ideological conflict. Their case requires the rejection of impartiality. They may cite actual lapses from impartiality to support this rejection, but such lapses are consistent, under the conditions stated in section 1.2, with upholding impartiality as a value of scientific practices. I consider upholding impartiality to be the hallmark of scientific inquiry. It is not "anti-science" to point out that impartiality is more fundamental than neutrality and autonomy, and that it may be realized where the other two are not. Rather, when this is grasped, we can come to recognize possibilities open to science, which have been obscured by the virtual uncontested privilege granted to materialist strategies. Questioning science is value free in the way that

I have, thus, can lead to criticizing, not the value and objectivity of scientific practices per se, but their status quo, and doing so for the sake of enhancing the manifestation of one the values (neutrality) that the tradition itself affirms and providing means to prevent lapses from another (impartiality).

The ground has now been cleared to permit conducting scientific inquiry in response to the question: How should scientific (systematic empirical) inquiry be conducted and institutionalized today so as to be of greatest service to furthering human well-being, in all of its dimensions, of as great a range of people as possible? This question need not be the source of "bias" and "distortion" entering into scientific inquiry; and, contrary to what the scientific mainstream may be inclined to react, it does not display misunderstanding of the nature of science or threaten its integrity. On the contrary, it can lead to identifying and challenging "bias" that may be present where a single kind of strategy predominates and where the grip on modern sensibilities (and their imaginative powers) of the modern valuation of control is strong; and responding to it is necessary in order that the projects of alternative value-outlooks can become informed by sound scientific knowledge. Since it underlies a rationale for strategic pluralism, it is actually more likely than research conducted virtually exclusively under materialist strategies to further neutrality and impartiality. The question posed above will become interpreted concretely as related questions are posed: How wide is the range of strategies that should be admitted into current scientific practices—in response to what value-outlooks, expressive of what conceptions of human well-being, and articulated by what range of social and political movements? How exactly do current scientific practices, and the ways in which they are institutionalized, contribute to furthering or diminishing human well-being? What anticipatory strategies might be explored more fully?

These questions cannot be answered a priori. Posing them, and figuring out the kinds of strategies under which inquiry can play a role in answering them, reflect the dialectical unity of the natural and social sciences proposed in the previous section. Consider, for example, the question about how widely, in response to what range of value-outlooks, we should multiply strategies for the sake of furthering human well-being. Producing an answer to it requires systematic empirical inquiry that is, in part, social scientific. This is because a fundamental dimension of human well-being is the exercise of cultivated, effective agency. Such agency is exercised when one acts in the important aspects of one's life, informed by one's beliefs, so that desires (objectives) that are expressive of a broad array of the values one holds are regularly satisfied (section 11.2; *SVF*: chapter 9; Lacey 2005b). Thus, answering the question requires understanding the range of values (personal, moral, social) that people desire to manifest in their lives, their variation with culture, class, and other factors, and what projects people have developed for the sake of furthering the exercise of their agency, and thus understanding the range of value-outlooks that frame conceptions of well-being that are espoused throughout the world. Unless we conduct this kind of social inquiry, we cannot discover what range of strategies need to be pursued in scientific inquiry for the sake of enhancing human well-being. This inquiry will help to locate limits to the neutrality of research conducted under materialist strategies and any other strategies that may be developed. It will also identify value-outlooks that may be interested in possibilities (e.g., those that would embody the values of popular participation—section 6.3; section 11.4), the exploration of which requires alternative strategies. Once the range of these value-outlooks has been identified, we can explore the set of strategies that may be desirable, in light of the question, for shaping research in current social circumstances. Given limited resources, however, it may not be possible to conduct research to the desirable extent under all the identified strategies. I think that it follows that it should be a matter of democratic deliberation and enactment that determines under which strategies research takes place and how priorities are set among them; certainly there is nothing about the expertise cultivated by professional scientific practitioners working under any particular strategy that prepares them for making these choices. It remains a major question to address the issue of how such democratic decision-making could happen in a reliable and effective manner.[3]

Posing the above questions has not usually been considered a central task of scientific practices and they get little consideration in the tradition of philosophy of science. Partly this is because mainstream science has tended to take for granted the presuppositions that I have diagnosed of the modern valuation of control, and so it assumes that, for the most part and in the long run, scientific knowledge gained under materialist strategies serves human utility well. But, the mainstream tends to see utility as a consequence of gaining such scientific knowledge; it does not consider the values covered by "utility" to have methodological impact at the moment of the adoption of materialist strategies. The question assumes importance only when the modern valuation of control is contested.

Things are changing. Lately, several philosophers of science have posed related questions and devoted considerable attention to them. Consider John Dupré:

> The most important topic in the philosophy of science [is] the relation of science to human values. What contributions can or should science make to human well-being? . . . If there is one conclusion of overriding importance to be drawn from the increasing realization in recent times that science is a human product, it is that, like other human products, the only way it can ultimately be evaluated is in terms of whether it contributes to the thriving of sentient beings in the universe. (Dupré 1993: 244, 264)

And Philip Kitcher:

> Reflective people . . . want to know whether research in various areas is skewed by the values of particular groups and, at the broadest level, how science bears on hu-

man flourishing. . . . It has been obvious for about half a century that research yielding epistemic benefits may have damaging consequences for the individual or even the whole species. Philosophical stories about science have been narrowly focused on the epistemic. Faced with lines of research that have the capacity to alter the environment in radical ways, to transform our self-understanding, and to interact with a variety of social institutions and social prejudices to affect human lives, there is a much larger problem of understanding just how the sciences bear on human flourishing. (Kitcher 1998: 46)

[And] To claim . . . that the sciences achieve certain epistemic goals that we rightly prize is not enough—for the practice of science might be disadvantageous to human well-being in more direct practical ways. A convincing account of practical progress will depend ultimately on articulating an ideal of human flourishing against which we can appraise the various strategies for doing science. The extreme positions are clear . . . science, as practiced, is a terrible thing, and that human beings want none of it; . . . science, as we have fashioned it, is already perfect. Neither extreme is likely to be right. . . . Given a clear view of the epistemic achievements and prospects of science, how should we modify the institution so as to enhance human well-being? (Kitcher 1993: 391)[4]

And, in a different tone, Paul Feyerabend, who has criticized science in the light of "humanitarian" and "democratic" considerations, and who has repeatedly suggested (Feyerabend 1993, 1999) that following a single approach to science (conducted under materialist strategies?) threatens human well-being (section 4.3):

Being related to each other in lawful ways, [the entities unearthed by science] can be manipulated or predicted by using these laws. There can be new combinations of them and new entities may in this way arise at the phenomenological level. But these entities are important only if the resulting world is pleasant to live in, and if the gains of manipulation more than compensate for the losses entailed by the losses of unscientific layers. The objection that the entities and laws that connect them are "real" and that we must adapt to them, no matter how dismal the consequences, has no weight. (Feyerabend 1999: 12)

Each of these philosophers introduces different emphases and idioms, all of which (in turn) are different from mine. I do not want to convey the misleading impression that a common vision about how science ought to be conducted has emerged. All concur, however, that questions about the impact of science on human well-being, and on how science ought to be practiced in the light of this impact, need to be addressed.

The questions stated above can no longer be brushed aside on the ground that values have no proper place in setting the direction of scientific research. Often, one way or the other, some values can be expected to play such a role, ensuring that understanding gained under a strategy is primed to be significant for value-outlooks containing these values. Better that the role played be a consequence of discussion and deliberation, rather than a covert one played in subtle dialectical interaction with the reigning values of the age. Of course, it is important that the role be played at the right (logical) moment, that of adoption of strategy, and never at the expense of impartiality. Recognizing this should not obscure there are rich dialectical interactions among the questions: "How to conduct scientific research?" "How to structure society?" and "How to further human well-being?" Science may be appraised, not only for the cognitive value of its theoretical products, but also (without threatening this) for its contribution to social justice and human well-being. This book makes a modest contribution to showing how this can be so.

NOTES

1. Some recent reactions of biotechnology researchers to criticism have been of this kind (Borlaug 2000; Potrykus 2001). Ironically they see threats to the "autonomy" of science coming more from their critics than from the corporate sponsors of much of their research (see also section 10.6).

2. This is an instance of a more general phenomenon, that there is typically a "gap" between the values that people articulate and those manifested in their behavior. In *SVF* chapter 2, I have maintained that holding values authentically involves commitment to reducing such gaps (see also section 3.1 and section 6.3).

3. Kitcher (2001), Longino (2002), and Proctor (1991) have made different proposals about how scientific research should be institutionalized in a democratic setting (cf. 10.4).

4. Kitcher has also discussed in detail how recent developments of biotechnology in medicine contribute towards enhancing or diminishing the "quality of life" (Kitcher 1997), and more recently (Kitcher 2001) he has analyzed the "significance" of scientific results as having both cognitive and social-value dimensions. Some recent work in the social studies of science might be considered to cast light on many of these questions though much of it tends to ignore the separation of moments (strategy adoption/theory choice) that I have emphasized. Kitcher is sensitive to this kind of separation, as also is Longino (2002).

Chapter Three

The Distinction between Cognitive and Social Values

I proposed (chapter 1) that what marks inquiry as "scientific" is commitment to impartiality, rather than being conducted under materialist strategies. Thus I take any form of systematic empirical inquiry, regardless of the strategy under which it is conducted, to be an instance of scientific inquiry provided that its results are held to accord with impartiality. Impartiality presupposes that there is a distinction between cognitive and social (moral, cultural, etc.) values. This distinction has recently been questioned and its significance for scientific methodology, challenged (Longino 1996; Machamer and Douglas 1999; Douglas 2000; see *SVF*: chapter 9; Lacey 1999). Prima facie there is a distinction between cognitive and social values that may be put starkly as follows: Cognitive values are characteristics that scientific theories and hypotheses should have for the sake of expressing understanding well; or, as Laudan puts it, they are attributes that "represent properties of theories which we deem to be constitutive of 'good' theories" (Laudan 1984: xii); whereas social values designate characteristics deemed constitutive of a "good" society.

Is this a significant distinction? Does it serve to illuminate key features of scientific knowledge and practices? The tradition of modern science has normally answered with a resounding "yes," and this constitutes part of the grounds for adherence to the view that science is value free. The integrity, legitimacy, prestige, and alleged universal value of science have often been tied to "science is value free" being highly manifested (with a trajectory of higher manifestation) in the practices of science—for, it is said, it is science conducted in such practices that has enabled the technological applications that have so transformed the world in recent times. I have questioned "science is value free" (in section 1.3), rejecting one of its components, autonomy, and pointing to ambiguities in another, neutrality. Nevertheless, I too think that there is a significant distinction between cognitive and social values, as I will argue in detail in this chapter. The possibility of accepting theories in accordance with impartiality—and thus of having sound scientific knowledge—depends on it. At the moment of accepting or rejecting theories (as required by impartiality), cognitive, but not social, values have an indispensable role. But this leaves an important role for social values at other moments, including those of adopting strategies for research and of applying scientific knowledge. The distinction is crucial for properly interpreting the results of scientific research, and for opening up reflection on how neutrality might be defended as a value of scientific practices at a time when much of its mainstream is becoming increasingly subordinate to the dominance of "global" capitalism (chapter 2; section 2.1). My defense of the distinction, unlike the traditional viewpoint, does not support keeping social values entirely out of the core of scientific activity. While it does continue to uphold impartiality and so to permit no role for social values at the moment of decisive theory appraisal, it does permit a role for social values at the other core moment, adopting a research strategy.

I will use a number of abbreviations in presenting the (rather complicated) argument of this chapter in order to make the presentation more concise and its structure more perspicuous. For easy reference I have listed all the abbreviations (along with those used in other chapters) in the glossary.

3.1 VALUES, VALUE JUDGMENTS, AND VALUE-ASSESSING STATEMENTS

We hold various kinds of values: personal, moral, social, aesthetic, cognitive; they are held in more or less coherent and ordered value-outlooks in which they mutually reinforce one another (*SVF*: chapter 2; see also chapter 11). Let ø designate a particular kind of value, v some characteristic that can be manifested to a greater or lesser degree in ø, and X a person. Then

> X holds v as a ø-value, if and only if
> (1) X desires that v be highly manifested in ø;
> (2) X believes that the high manifestation of v in ø is partly constitutive of a 'good' ø; and
> (3) X is committed ceteris paribus to act to enhance or to maintain the degree of manifestation of v in ø.

In the case of social values, ø = society (social institutions, social structures), and a social value (sv)—for example, respect for human rights—is a feature whose high manifestation is valued in society; in the case of cognitive values, ø = accepted theory (hypothesis); a cognitive value (cv) is a feature whose high manifestation is valued in accepted theories (hypotheses), that is to say, in accepted bearers of understanding.[1]

"X holds the ø-value v" will be considered equivalent to "X makes the ø-value judgment, V, of valuing that v be a well-manifested feature of ø." (V represents the value judgment made when v is held as a value.) Basic value judgments are broadly of three types: "that v is a ø-value, a characteristic of a 'good' ø," "that v_1 ranks more highly (as a ø-value) than v_2," and "that ø manifests v to a 'sufficiently high' degree." There are also value judgments of the form: "that u has ø-value" or "that u is an object of ø-value," which are made on the basis of u's contributing to the manifestation of the ø-value. A methodological rule can be said to have cognitive value in virtue of its causal contribution towards generating and/or confirming theories that manifest the cv highly. On similar grounds, scientific practices organized so that they manifest certain sv (Longino 1990; 2002), or the cultivation of certain ethical virtues among scientists, may be said to have cognitive value. It is also likely to be a sv for X that the cv (X holds) be socially embodied: that there be social institutions (scientific ones) that nourish practices in which theories (more and more of them) highly manifesting these cv come to be accepted. Moreover, a theory (T), which manifests the cv highly, may also have social value (for X)—or significance (section 1.2)—in virtue of making a contribution to the manifestation of X's sv, on application or just by providing understanding of some phenomena. Thus, although literally a cv could not be identical to a sv, nevertheless "is an object of social value in the light of a set of sv" can designate a property of T. In affirming that cv and sv are distinct, I am committed to the claim: "There is a set of cv, and "is an object of social value in the light of a set of sv" does not define one of them."

There are close links between the values X holds and those embodied in the institutions in which X participates; and, at least sometimes, that these are embodied in the institutions explains why X holds these values rather than others. (That the modern valuation of control is held widely may be explained by its deep embodiment in leading social institutions.) For those who consider values to be subjective preferences, that explanation is the end of the story. On my account values are not reducible to subjective preferences. Another distinction needs to be brought into play, one between X's value judgment V that v is a ø-value and a statement of fact about the degree of manifestation of v in ø.[2] Values are features that ø may or may not exhibit, typically features that ø manifests (exhibits, possesses) to a greater or lesser degree. (Respect for human rights may be manifested in a society and explanatory power may be manifested in a theory to a greater or lesser degree.) To what degree ø manifests v is a matter of fact. I will call statements expressing such matters of fact "value-assessing statements." They are not value judgments, although statements expressing that v is manifested in ø to a "good enough" degree are; they are hypotheses that can be part of theories investigated in the social sciences, and that can be appraised (under an appropriate strategy) in the light of available data and the cv. There are kinds of statements that are frequently used that simultaneously express value-assessment statements and value judgments. Consider: "The United States respects human rights." This expresses both: "Respect for human rights is manifested in such and such a way and to such and such a degree in the United States" and "This way and degree are adequate" ("'good enough' for now" or, in certain contexts, exemplary). This is an instance of a statement deploying "thick ethical terms" ("'thicker' or more specific ethical notions . . . such as 'treachery,' and 'promise' and 'brutality' and 'courage,' which seem to express a union of fact and value" [Williams 1985: 129]). I suspect that all sentences containing thick ethical terms are best understood as conjunctions of value-assessment statements and value judgments. Later (section 3.4), when discussing questions about the legitimacy of technological innovations, I consider statements—about side effects and alternatives—that we "endorse" rather than "accept." These statements display a similar logic to those that deploy thick ethical terms.

There are deep interconnections between value judgments and value-assessing statements. In the first place, it is unintelligible that one make V and not be able to appraise and affirm value-assessing statements of the kinds: "ø manifests v to a greater degree at time t_1 than at t_2," "an institution embodies v to a greater degree at time t_1 than at t_2, or to a greater degree than another institution does," and related ones. That is because, without such assessments, X could not know whether the desire of item (1) were satisfied or not, so that V could not lead coherently to any action (as required by item (3)), but would remain merely a verbal articulation devoid of behavioral significance.

Secondly, in view of item (2), it is a presupposition for X to make V that X believes that it is possible that ø manifest v highly (or, if the current manifestation is low, to a higher degree than it does now); for it is unintelligible to affirm: "v is feature of a 'good' ø, but v cannot be highly manifested in ø." Respecting human rights is not an sv unless respect for human rights can be highly manifested throughout one's society. The total elimination (as distinct from the drastic reduction) of injustice is not an sv because it is impossible to totally eradicate injustice. Predictive power is not a cv unless predictions can be generated from theories. Certainty is not a cognitive value because theories that are known with certainty cannot be generated with the scientific methods that we deploy. Evidence for or against relevant possibility statements may be sought for and obtained in scientific investigations; it will be based in large part on confirmed value-assessing statements. Evidence that counts decisively against the possibility statement constitutes ceteris paribus a reason to reject V. Value judgments may thus be affected, in logically permissible ways, by the outcomes of scientific inquiry (Lacey 1997).

Thirdly, in item (2) of the schema, "good" is functioning as a sort of placeholder. In practice, however, "good" tends to carry with it an ideal of ø—a general, fundamental, comprehensive aim, goal, or rationale for ø of which the ø-values are constitutive (or to which they are subordinate). Such ideals may not be well articulated. In any case, they always remain open to further articulation and, in the process, no doubt, to development and revision. A good deal of value discourse is marred by inabilities to communicate that are consequences of different but poorly articulated ideals being in play (section 6.3). Much of it is also marred by gaps, of which people may or may not be aware, between what they say their values are and what their behavior shows; and the more powerful one's interlocutor, the less likely it is that he will be moved by arguments and evidence (e.g., pertaining to the factual presuppositions of value judgments) that may call in to question the values that his behavior shows to be in play. When we take ideals of ø into account (next section), the interconnections of value judgments with value-assessing (and other factual) statements become more extensive.

3.2 SOCIAL IDEALS AND COGNITIVE IDEALS (THE AIM OF SCIENCE)

Consider social values. X might articulate (as I do—chapter 11) as the fundamental ideal or rationale for society that it provide structures that are sufficient to enable all people normally to live in ways that manifest values that, when woven into an entire life, generate an experience of well-being (fulfillment, flourishing). This ideal, in turn, needs to be complemented by an account of human well-being (a view of human nature) and of how it is intertwined with X's ethical ideals. Then v is an sv for X only if X believes that the manifestation of v is constitutive of such structures, or of institutions or movements that aim to bring them about (chapter 11).

Generally, regardless of the social ideal that is articulated, holding v as an sv presupposes:

a) Higher manifestation of v contributes to fuller realization of the ideal;
b) Higher manifestation of v, and acting to bring it about, do not undermine higher manifestations of other held sv; and
c) There is not another characteristic, which cannot be highly manifested in the same structures as v, whose higher manifestation would contribute more towards the realization of the ideal.

These three presuppositions are in addition to the earlier one:

d) It is possible for v to be more highly manifested (or for its current high manifestation to be maintained) in the society;

and this will be true only if:

e) In the society conditions are provided (or can be created) that ensure the availability of objects of social value, which are such in virtue of their causal contribution towards the higher (or maintained) manifestation of v.

All of these presuppositions are open to input from empirical investigation. While they do not entail V (the value judgment made when v is held as a value), positive support for them provides good backing for one to make V, and evidence against them points to the (rational) need either to revise one's sv or one's articulation of the social ideal. That is significant, even if reaching agreement about the ideal may appear to be out of reach. Suppose, for example, that strong empirical evidence were provided that a valued economic arrangement is a significant causal factor in the perpetuation of widespread hunger, thus showing a causal incompatibility between two plausible sv. Something has to give. We are not stuck with across-the-board intractable disagreement. Dialectically, empirical investigation can be crucial in disputes about sv. (This is exemplified throughout part II; see also chapter 11.)

The Aim of Scientific Practices

Now turn to cognitive values. A cv is a feature whose manifestation is valued in accepted theories (hypotheses); it is a characteristic of "good" accepted theories. X, qua participant in scientific practices in which theories are entertained, pursued, developed, revised and appraised, subscribes to an ideal of what makes an accepted theory "good", that is, a general, fundamental, comprehensive aim, goal, or rationale for theories, of which cv are constitutive. While individuals may differ in their judgments about what is constitutive of a "good" accepted theory (i.e., about the cv), so long as there is agreement that theories are intended to be bearers of understanding and knowledge about phenomena that can (often) be expected to be applied successfully in social practices, the differences will be considered disagreements in need of resolution. Whatever one might maintain about the "subjectivity" of other kinds of values (my account of values maintains some elements of subjectivity), there is little room for it in connection with cv (Scriven 1974). Since theories are products of scientific practices, I find it convenient to locate the ideal attributed to theories against the background of the ideal (X's) of scientific practices.

I will attribute to X the ideal or aim of scientific practices:

AS: (i) to generate and consolidate theories that express empirically grounded and well-confirmed understanding of phenomena,
(ii) of increasingly greater ranges of phenomena,

(iii) such that no phenomena of significance in human experience or practical social life—and generally no propositions about phenomena—are (in principle) excluded from the compass of scientific inquiry, and
(iv) with a view (where appropriate) to the practical application of the knowledge represented in well-confirmed theories.

I take understanding to include descriptions that characterize what the phenomena (things) are, proposals about why they are the way they are, encapsulations of the possibilities (including hitherto unactualized ones) that they allow in virtue of their own underlying powers and the interactions into which they may enter, and anticipations of how to attempt to actualize these possibilities (*SVF*: chapter 5). I have stated the aim so as to encompass all inquiries that are called "sciences" (including social sciences) and those that bear close affinities with them. Recall that I include under "science" all forms of systematic empirical inquiry, because I want neither to rule out by definitional fiat, nor to assume a priori, that forms of knowledge that are in continuity with traditional forms of knowledge may have comparable cognitive (epistemic) status to those of modern science. Furthermore, I want to include within the compass of scientific inquiry, not only investigations that lead to efficacious applications, but also those pertinent to appraising the presuppositions of the legitimacy of applications (part II). Then, I do not restrict what counts as a theory to that which has mathematico-deductive structure or which contains representations of laws, but also include any reasonably systematic (perhaps richly descriptive or narrative) structure that expresses understanding of some domain of phenomena.[3]

It follows that v is a cv only if it is a characteristic of theories, the sound acceptance of which furthers AS(i); thus only if it is constitutive of a theory's expressing empirically grounded, sound understanding of a range of phenomena. The cv (more accurately, gaining theories manifesting them) are constitutive of the cognitive aims of scientific practices. Minimally, then, cv must in fact—as revealed in interpretive studies that, among other things, explain why the currently articulated cv have displaced those articulated at earlier historical times (Laudan 1984)—play the role of cognitive aims in the tradition (or traditions) of scientific inquiry and have become manifested in theories whose acceptance is currently settled.[4] Moreover, if a robust notion of understanding is to be maintained, there should be compelling reasons why other candidates that have been proposed, for example, "being an object of social value in the light of some sv", are not included among the cv.

As already noted, v is a cv only if it is possible that theories manifest v highly, and that is the case only if there are available methodological rules that enable its manifestation to be furthered. There are also counterparts of the presuppositions of holding sv, (a)–(c), that the aim, AS, renders intelligible. In the present argument, however, I will only draw upon the counterpart of presupposition (b): there are no undesirable side effects in scientific inquiry of holding v as a cv:

B: (a) Gaining the higher manifestation of v in a theory is not incompatible with other cv gaining increased manifestation in that theory; and (b) requiring that v be manifested in an accepted theory does not inhibit entertaining (as potentially worthy of acceptance) theories that might manifest some of the other cv well.

Nothing said here about conditions on cv—except perhaps including (b) in B—goes beyond Laudan's "reticulated model" (Laudan 1984), and everything I have said is influenced by his model.

B suffices to eliminate a large range of candidate cv. The range includes commonly identified "outside interferences" that have been considered to threaten the autonomy of the conduct of science: accepted by consensus, popularly believed, consistency with the presuppositions of particular social values, consistency with biblical interpretations or with the tenets of dialectical materialism. Consistency with biblical interpretations, for example, clashes with empirical adequacy, keeping ad hoc hypotheses to a minimum, and power to encapsulate the possibilities permitted by actual phenomena (Mariconda and Lacey 2001). The elimination is not achieved a priori but it is empirically grounded and enacted dialectically. Consider the candidate that it is important for me to eliminate: being an object of social value in the light of specified sv (which I will call OSV). OSV is not a cv. Scientific inquiry at various times has deployed varieties of it at the cost of inhibiting the entertainment of theories, which at other times have been shown to manifest the virtually unanimously held cv well. These cv include: empirical adequacy, explanatory power, consilience, keeping ad hoc hypotheses to a minimum, power to encapsulate possibilities of phenomena, and containing interpretive resources that enable the explanation of the success and failures of earlier theories (*SVF*: chapter 3). It might be proposed, nevertheless, that there may be a particular set of sv for which this kind of inhibition does (would) not occur. That could only be settled empirically.

Understanding and Utility

Since the outset of modern science there has been controversy about the relationship of understanding and utility. A relationship is acknowledged in the statement of AS. Item (iv) is intended to permit matters connected with practical application (or service to interests shaped by particular sv) to play a role connected with decisions about focus or priorities of inquiry, while maintaining a balance with (iii). There has sometimes been a tendency to build utility into the aims of science; and even to modify (iv) to specify that scientific theories should have the property of OSV for certain sv, for example, the modern valuation of control. (Baconian influence has never been absent from the modern scientific tradition.) On most views, however, understanding is a prerequisite for utility; a necessary condition for a theory's having the property OSV is that it manifest the cv to a high degree, so that from the aim of utility we do not get criteria that can be brought to bear on what constitutes sound understanding. Utility leads us to emphasize certain lines of investigation, requiring that possibilities, such as enhancing our capability to control natural objects, be addressed, and that other lines be neglected because they cannot be expected to give rise to theories that are likely to be objects of social value in the light of held sv. When inquiry is conducted in this way all theories that come to be accepted will have the property of OSV for these sv. These theories have OSV because, from the outset of the research process, only theories that would have it, if they were to become soundly accepted, are even provisionally entertained. That a theory actually (and not just potentially) may have the property of utility follows from its being accepted (of a certain domain of phenomena) in view of its manifesting the cv to a high degree.

The role of utility (OSV) in the selection of theories (when it plays a role) is at the moment where the kinds of theories to be provisionally entertained and pursued are chosen, not where choice among provisionally entertained and developed theories is made (see section 3.4 below). What is discovered in the course of investigation is that the theories manifest the cv highly; that they also manifest OSV is simply a consequence of constraints imposed on the inquiry at the outset. If no theories were consolidated under these constraints, we would discover that there are no theories manifesting OSV. OSV does not play a role alongside the cv at the moment where theories come to be accepted (section 3.4). Even if the body of scientists were interested only in investigating useful theories, it would not follow that all theories that can express understanding must manifest OSV. One can build utility (as distinct from the much weaker and more abstract (iv) of my formulation) into the aims of science (and include OSV among the cv) only by prima facie clashing with AS. I hasten to add that concluding that OSV is not a cv might be quite trivial, if the scientific tradition had not made available a large body of scientific understanding, vindicated in virtue of being expressed in theories that have been soundly accepted of certain domains of phenomena, with respect to which they manifest the cv to the highest degree available. "Utility" has been well served by this understanding.

3.3 STRATEGIES, METHODOLOGICAL RULES, METAPHYSICS, AND SOCIAL VALUES

One might respond: Even if, for any sv, OSV is not a cv, nevertheless there may be a set of sv such that the methodological rule (MR-OSV) is an object of high cognitive value.

MR-OSV: Entertain and pursue only theories that are constrained so that, if soundly accepted, they also manifest OSV (being an object of social value in the light of the specified sv).

If the historical record were to show that following this methodological rule furthers the realization of the aim of science, AS, then the distinction between cv and sv would be of less importance.

In rebutting this response, I note that AS pulls us in two directions, which in actual scientific practice can be in tension. One pull is towards putting research effort where we can expect to have readily recognizable success quickly and efficiently. The other is towards engaging in research on phenomena that have been underinvestigated where, since scientific research builds upon its past achievements, theories may appear to be underdeveloped, lacking in internal sophistication, and not very general in scope. Later I will argue that it is important for science, qua worldwide institution, to react to both pulls. I think that the first pull has dominated modern science, however, and that it has indeed led to a methodological rule of the kind just stated being taken as an object of high cognitive value. Furthermore, adopting this methodological rule has enabled some reactions to the other pull, as reflected in the regular emergence of new branches of modern science, such as, recently, cognitive science (Lacey 2003b). Nevertheless what is "left out" is of great importance, as I will argue below.

Materialist Strategies and the Methodological Rule to Conduct Research under Them

Before stating the methodological rule of the type MR-OSV that I have in mind, I will first introduce another methodological rule, MR-M, which functions in close concert with it, but which does not, on the face of it, link the pursuit of science with any social values.

MR-M: Entertain and pursue only theories that instantiate general principles of materialist metaphysics about the constitution and mode of operation of the world.

I take materialist metaphysics to affirm that the world (phenomena and the possibilities they permit) can be adequately represented with categories that may be deployed under appropriate varieties of materialist strategies (abbreviated S_M); so that adopting MR-M is tantamount to adopting the methodological rule, MR-SM:

MR-SM: Entertain and pursue only theories that meet the constraints of S_M, that is, that represent phenomena and encapsulate possibilities in terms that display their lawfulness, and thus usually in terms of their being generated or generable from law and/or underlying structure, process, and interaction, dissociating from any place they may have in relation to social arrangements, human lives and experience, from any link with value, and from whatever social, human, and ecological possibilities that may also be open to them; and (reciprocally): Select and seek out empirical data expressed using descriptive categories that are generally quantitative, applicable in virtue of measurement and instrumental and experimental operations—so that they may be brought to bear evidentially on theories entertained and pursued under the constraints.

On this account, materialist metaphysics does not entail physicalism, any form of reductionism, or determinism, and its concrete content is not fixed. Its articulations develop with the developments and changes of S_M—and thus with decisions (which can readily be shown to fit Laudan's "reticulated model") about which versions of MR-SM to follow at a given time in a given field—and so they are not tied to any particular versions of S_M, certainly not to surpassed mechanistic versions.

Although modern science has been conducted virtually exclusively under varieties of S_M, I have maintained that they are one (albeit a very special one) among a plurality of kinds of strategies that might be adopted in systematic empirical inquiry. That any one of these kinds should be prioritized in inquiry does not follow from AS. Indeed, neither direction to research, nor definition of what counts as worthwhile or significant research, follows from AS (section 1.3); these come with the adoption of a strategy. Recall that the principal roles of a strategy are to constrain the kinds of theories (hypotheses) that may be entertained in a given domain of inquiry (so as to enable investigation) and the categories that they may deploy—and thus to specify the kinds of possibilities that may be explored in the course of the inquiry—and to select the relevant kinds of empirical data to procure and the appropriate descriptive categories to use for making observational reports. Different classes of possibilities may require different strategies for their investigation.

Since, under S_M, intentional and value-laden categories are deliberately excluded from use in the formulation of theories, hypotheses, and data, where S_M are deployed there cannot be any value judgments among the formal entailments of theories and hypotheses. Following MR-SM thus suffices to ensure that the first presupposition of neutrality, that theories have no value judgments among their entailments (section 1.2) is satisfied. This is a feature of the design of S_M. There can be no doubt that conducting research under them has been fruitful and that it has furthered AS to a remarkable extent; there are plenty of theories, expressing understanding of an ever increasing number and variety of phenomena, developed under S_M that manifest the cv to a high degree. Moreover, S_M have proved to be highly adaptable, and new varieties of them have been created as research has unfolded: varieties expressing mechanism, lawfulness expressed mathematically, various forms of mathematical laws (presupposing Newtonian space and time and relativistic space-time; deterministic and probabilistic; with and without physicalistic reductionism; functional, and compositional), computer modelling, molecular and atomic structures, etc. Thus, there can be little doubt that research that follows MR-SM is indefinitely extendable and we can expect that AS would continue to be furthered by following it. Minimally, then, MR-SM' has high cognitive value, where MR-SM', "Entertain and pursue theories that are developed under S_M," differs from MR-SM by dropping the "only" in the latter's formulation.

Bounds of Research Conducted under Materialist Strategies

It does not follow, however, that, even in principle, all phenomena may be understood within the categories that MR-SM allows. That is because (elaborating the line of argument of section 1.3) under S_M phenomena are investigated in dissociation from any context of value, and so any possibilities they may derive from such contexts are not addressed. Routine explanations of human action (including of actions involved in scientific research and applied practices), and attempts to anticipate it, eschew such dissociation. They deploy categories (e.g., intentional and value ones, including those used in value-assessing statements) inadmissible under S_M, and theories containing them may manifest the cv highly (Lacey and Schwartz 1996; Lacey 2003b).

Currently, while there is no compelling evidence that these theories are reducible to or replaceable by theories constructed under S_M, varieties of S_M have been adopted in the behavioral, cognitive, and neurological sciences anticipating that reductions or replacements will be found. This is an instance of how science, following MR-SM, is responsive to the pull (the second pull of AS) to grasp new kinds of phenomena. While theories in these fields have been soundly accepted of some (mainly experimental) domains of phenomena, that they may be extended into characteristic phenomena of human agency remains simply an anticipation with slender empirical warrant (but often fostered by metaphysical materialist theories of the mind). The conjecture that adequate understanding of human agency can be produced under S_M is worth exploring and a priori I do not rule out its eventual confirmation, but it can only be tested rigorously against the products of research conducted under strategies (call them "intentional strategies") that do not make the dissociations made by S_M. To limit research to that conducted under S_M would inhibit the exploration of theories under which, for example, explanatory adequacy might become highly manifested. Then, furthering of AS—for example, in connection with the investigation of value phenomena—would require deployment of strategies (e.g., intentional ones) under which such theories could be entertained and appraised (*SVF*: chapter 9; Lacey 2003b; chapter 11).

In the light of such considerations, I asked (in section 1.3): Why has research conducted under S_M been so dominant—so much so that adopting them has often been seen as being essential to scientific research, and research that does not adopt them (as in some social sciences or in agroecology) as not quite "scientific"? What explains it? And what rational justification can be given for it? Appeal to materialist metaphysics does not justify it (*SVF*: chapter 6) for it lacks the necessary cognitive (empirical) credentials. Could it be that the first pull of AS is so strong that, so long as S_M continues to expand into new fields, scientists are content to limit themselves to gaining the understanding they are confident of gaining now? And to leave it to the future (when one may anticipate that there will be still more sophisticated versions of S_M available) to address phenomena that are unexplored under these strategies now? Clearly there is something of explanatory import to this suggestion, but it does not supplant the explanation I proposed in section 1.3 (the link with the modern valuation of control) and it cannot be sustained as a justification when one takes into account the application of scientific theories.

The Importance of Considering Applications

I have already indicated that theories may be appraised not only for their cognitive value but also for their social value. Can they, on application, inform projects valued in view of specified sv? The traditional answer is that any theory that is applicable at all can (in principle) inform projects of any viable sv, and so those of the specified ones. Since I find this wildly implausible (*SVF*: chapter 10), I have recast neutrality to affirm that the body of accepted theories are evenhanded on application across value-outlooks; for a given sv some soundly accepted theories may be significant for it and some not, but overall it is (in principle) as well served by applications of scientific theories as any other. This cannot be sustained if research is principally confined under S_M. As pointed out above, theories developed under S_M indeed satisfy a presupposition of neutrality, but they need not thereby manifest neutrality. To illustrate, transgenic plants are embodiments of soundly accepted theoretical knowledge developed under versions (biotechnological) of S_M (abbreviated, S_{BT}). As technological objects, however, they have no significant role in the projects of those who aim to cultivate productive, sustainable agroecosystems in which simultaneously biodiversity is protected and local community empowerment is furthered. (These values are included in those of "popular participation" (section 6.3)—abbreviated V_{PP}.) Hence transgenics have little social value for the many rural grassroots movements throughout Latin America (and elsewhere) that hold these values. These applications of knowledge gained under S_{BT} do not display evenhandedness. Nevertheless, projects aiming to further V_{PP} do not lack scientific input. Theories that inform them are consolidated by research conducted under agroecological strategies (S_{AE}), strategies under which a multiplicity of variables (concerning crop production, ecological soundness, biodiversity, and local community well-being and agency) are investigated simultaneously and interactively (section 5.4).

What is included in the range of sv, for which a theory may have social value, is a matter of fact, open to empirical inquiry. While I believe that many theories confirmed under S_M tend to have (varying) measures of social value for quite a wide array of value-outlooks, taken as a whole, when S_M are pursued virtually exclusively (as detailed in section 1.3), they have social value generally and especially in the light of value-outlooks that give primacy to the modern valuation of control (abbreviated V_{MC}).

Materialist Strategies and their Links with the Modern Valuation of Control

This fact contributes to explain the predominance of research conducted under S_M and the kind of epistemic privilege attributed to theories soundly accepted under them, for V_{MC} is widely held in modern societies and reinforced by its relationships with other values that are highly manifested in powerful contemporary social institutions (generally at the present time linked with capital, the market, and the military—abbreviated as $V_{C\&M}$). The explanation I offered, however, is not that S_M are adopted for the sake of generating applications that further interests cultivated by V_{MC}, but rather it draws upon there being mutually reinforcing relations between holding V_{MC} and adopting S_M. In virtue of these mutually reinforcing relations, theories soundly accepted under S_M tend to be objects of social value especially in those institutions where V_{MC} is embodied highly. (Adapting my earlier notation, these theories manifest OVMC; and often, with respect to other sv, they do not manifest OSV significantly.) It is a condition of their manifesting OVMC highly and a ground for expecting the efficacy of their applications, of course, that their cognitive value is confirmed under S_M. That S_M have become the predominantly used research strategies is explained by the fact that theories, which become soundly accepted under them, also manifest OVMC, and thus effectively are products of following the methodological rule (an instance of MR–OSV):

MR–OVMC: Entertain and pursue only (virtually exclusively) theories that are constrained so that, if soundly accepted, they also manifest OVMC.

MR–OVMC is the methodological rule, widely taken to be an object of high cognitive value among the practitioners of modern science, to which I referred at the outset of this section (section 3.3).

The link between S_M and V_{MC} not only explains, but also provides the rational grounds (if there are any) for the predominance given to research under S_M.

Thus, I suggest, adopting MR–M is tantamount to adopting not only MR–SM, but also MR–OVMC. According primacy to research conducted under S_M is not derived simply from commitment to furthering AS, although research conducted in this way has enabled AS to be furthered to a remarkable degree; and, at the same time, its products have informed the processes that have so entrenched the embodiment of V_{MC} in the leading contemporary socioeconomic institutions. Rationalizing the predominance of research conducted under S_M depends upon the rational upholding of V_{MC} (reinforced by commitment to the highly manifested $V_{C\&M}$ whose augmented manifestation depends on that of V_{MC}—section 10.6). Many people, perhaps awed by the trajectory of the increasing manifestation of V_{MC} and of the power that nurtures the social values with which it is linked, instinctively assume that V_{MC} contains a set of universal values (Presupposition (c) of V_{MC}—section 1.1), so that they recognize no difference between being applicable in service to this set of values and to all viable value-outlooks. Both V_{MC} and materialist metaphysics are deeply rooted in the consciousness of educated people in the advanced industrial nations and their allies in other nations, so much so that they find it hard to conceive that either one of them might be seriously questioned. Then, it may seem apparent that scientific research is identical to that conducted under S_M, and furthering AS reduces to carrying out research under S_M.

The Need for a Plurality of Strategies

In contrast, I pointed out above that sometimes, for example, concerning research in agriculture, there is a choice of strategies to be made (and the argument will be strengthened in section 5.4), for S_M lack the resources needed to explore certain classes of possibilities. In such situations, furtherance of AS would require that research be done, within the collective body of scientific institutions, under both S_M and competing strategies. When competing strategies are not pursued, I have just suggested, the key factor is upholding V_{MC}. Of course, for those who reject V_{MC}, this does not count as justification, and it poses no rational barrier for them to adopt different strategies that promise to provide knowledge and to identify possibilities that may inform projects with roots in their own values (e.g., adopting S_{AE} in view of their links with the values, V_{PP}.)[5]

Efficacy and Legitimacy

It is clear enough that, under S_M, possibilities are identified such that, when they are realized in technological objects or other interventions, the interests of V_{MC} and the sv linked with them ($V_{C\&M}$) are likely to be enhanced (though not always of these sv exclusively). Generally application presupposes efficacy: that a certain intervention or technological object with a specified design will work effectively, that it will perform as intended. Theories soundly accepted under S_M can be counted on to be an abundant font of efficacious applications of value in the light of V_{MC}, whereas theories developed under competing strategies, whatever efficacy they might engender, tend to further competing sv. The role of V_{MC} is even more far-reaching, however, since application involves accepting hypotheses not only about efficacy, but also about matters that underlie legitimacy.

Legitimacy involves consideration not only of the social value that will be directly furthered by an efficacious application, but also it requires the backing of the hypotheses:

No serious negative side effects (NSE): There are no effects—of significant magnitude, probability of occurrence, unmanageability—of negative social value caused by the application;

and often also of:

No "better" way (NBW): There is no other way, potentially of greater social value, to achieve the immediate goals of the application (or competing goals with greater social value).

The general discussion of legitimacy that follows will be exemplified, and developed in a much more nuanced way, in discussing the legitimacy of uses of transgenics throughout part II. In practice, the importance of NSE is generally recognized, and no application is introduced without some attention being paid to it.[6] Three things should be noted:

First, addressing NSE is usually assigned to studies of risk assessment. This involves making conjectures (based upon theoretical considerations or suggested by observations) about possible negatively valued side effects and then designing studies—in which specific problems, normally open to investigation under S_M, are posed—to determine their likelihood, manageability, etc. These are problems where answers are not settled simply by assessments of cognitive value; matters of ethical and social value are also essentially involved. That is because the empirical investigations of risk analysis presuppose answers to questions like: What is considered to be a serious risk? What are the mechanisms of risk? What is the relevant time frame for investigating risks? What are the evidential standards for judging that unmanageable risk is not present? What counts as sufficient evidence? Who should assume the "burden of proof"? And answers to these questions are all implicated in value judgments concerning the ethical stakes involved. Thence, any stance taken on NSE will involve the play of both cv and sv (and ethical values). (Below, section 3.4, I say that we are only able to "endorse" rather than "accept" hypotheses like NSE.) Furthermore, questions about the generality of proposed answers, and of unknown risks, are always hovering in the background; and this kind of risk assessment has no means to explore the effects of applications, qua objects of social value under specific socio-economic conditions (section 9.5).

Second, NSE is of negative existential logical form. Evidence for it is largely the failure to identify empirically well supported counter instances. (It may also gain evidential support from theories in which it may be embedded.) But absence of identified counter instances is not the same thing as the failure to identify them. Relevant failure is failure after having conducted appropriate investigation. Since NSE cannot be formulated using only the categories of S_M, the absence of systematic empirical inquiry that pursues theories in which it is contained may leave the attempt to identify counter instances a pretty haphazard affair. There is a techno-scientific mindset to defer considerations of risk until such time as actual harm being caused shows the risk to be serious (chapter 1: note 2). Where this mindset is operative, NSE tends to be presumed unless demonstrated otherwise (section 9.4). Third, where V_{MC} is held, and research conducted under S_M privileged, once standard risk assessment has been carried out, the "burden of proof" is placed squarely on critics. Yet, often, in view of the way in which science is institutionalized, the conditions are not available for them to assume the burden. The consequence is that NSE may go unchallenged even though, were the conditions available, a cognitively well-based critique might be able to be developed. NSE does get on the agenda, however, and standard risk assessments are the response; and in numerous cases that is quite adequate.

Sometimes when a technological innovation is introduced, NBW does not even make the agenda. Where a proposed alternative way is linked with sv that are in tension with V_{MC}, it tends to be dismissed (sometimes casually, sometimes for good reason); since V_{MC} tends to be taken as a set of universal values, the tension tends to be taken to be sufficient for rejecting these sv. When this happens, no serious consideration is given to whether conducting research under strategies linked with these sv might generate knowledge that would open up new possibilities that (unlike those identified under S_M) may serve to further interests fostered by the sv (section 5.4). This runs counter to commitment to neutrality. If one considers AS to express the aim of science, all of this line of reasoning is questioned. AS does not rest easily with the ready rejection of neutrality—even when that neutrality is subordinated only to values of such widespread acceptance as those included within V_{MC}. Neither neutrality nor serious assessment of NSE and NBW can be achieved without research being conducted under a variety of strategies, respectively linked with different sv. But, where V_{MC} is widely held, since the presumption in favor of the legitimacy of techno-scientific applications is so strong, relevant investigation of NSE and NBW is inhibited; and so their acceptance tends de facto to follow from this presumption rather than from empirical inquiry (contrary to impartiality).

76   Chapter Three

Impartiality, but Not Autonomy of Methodology

A methodological rule has cognitive value if it contributes to furthering AS. As pointed out above, MR–SM' clearly has cognitive value in connection with exploring many phenomena and possibilities. MR-SM, insofar as it encompasses MR-SM', contributes to furthering AS (i-ii) and (under a narrow definition) (iv), but insofar as it goes beyond it by virtue of the "only," it tends to undermine AS(iii). On the other hand, adopting S_{AE} furthers (iii); and also (i) and (ii), even if quantitatively to a much smaller extent than adopting S_M. There are certain classes of possibilities that, so far as we know at the present time and can anticipate for the foreseeable future, if they are genuinely realizable, cannot be identified by research conducted under S_M—for example, potential objects of social value given specified sv (e.g., V_{PP}), and those that might be identified under another strategy (e.g., S_{AE}). Nothing in my argument suggests that S_M should be dropped from scientific practices; rather, for the sake of furthering AS, they should be complemented by other strategies. I do not know what might be the range of other strategies that would be useful. In efforts to further AS I suspect that S_M will always have a special place (*SVF*: chapter 10), for strategies like S_{AE} draw upon the results of research conducted under S_M in numerous indispensable ways (section 5.4). Such strategies should not be seen as complete alternatives to S_M, but more as an interlocking set of local approaches, each of which draws upon the results of S_M where convenient. It is holding some sv, which conflict with V_{MC}, that makes it especially interesting to adopt these strategies, just as (if I am right) it is holding V_{MC} that grounds adopting effectively only S_M. Conflict about sv thus impinges on what methodological rules to follow, and thus autonomy of methodology (section 1.2) cannot be upheld. But impartiality, as an approachable ideal of accepted theories, is left untouched. Many theories developed by following MR-SM have been soundly accepted of wide-ranging classes of phenomena in virtue of manifesting the cv to a high degree. This is unaffected by the virtual equivalence of MR-SM and MR-OVMC. A similar logic applies where other strategies are chosen.

3.4 THREE MOMENTS OF SCIENTIFIC ACTIVITY

Out of this analysis, a model of scientific activity emerges (expanding the one sketched in section 1.3) in which it is useful to distinguish (analytically, not temporally) three moments of scientific activity at which decisions have to be made: (i) adopting a strategy, (ii) accepting (rejecting) theories, and (iii) applying scientific knowledge.

To accept a theory (T) is to deem that T needs no further testing and that it may be taken as a given in ongoing research and social practice. According to impartiality, T is soundly accepted of a specified domain of phenomena if and only if it manifests the cv to a high degree and if, given current "standards" for "measuring" the degree of manifestation of cv, there is no plausible prospect of gaining a higher degree (impartiality—section 1.2) Given AS as the aim of science, there is no rationally salient role for sv at the second moment; the fact that T may manifest some OSV highly counts rationally neither for nor against its sound acceptance.

At the first and third moments, however, sv have legitimate and often indispensable rationally relevant roles. At the third, obviously an application is made because it is intended to serve specific interests, and thus to further the manifestation of specific sv, and judgments of its legitimacy depend upon a multiplicity of value judgments. At the first, a strategy may (as we have seen) be adopted—subject, in the long run, to research under it being fruitful in generating theories that become soundly accepted—in view of mutually reinforcing relations between adopting it and holding certain sv, and the interest in furthering those values. (Sometimes it may be adopted for other reasons.) Adopting a strategy defines the kinds of possibilities that may be identified in research, (in important cases) possibilities that, if identified and actualized in applications, would serve interests cultivated under the sv linked with adopting the strategy. Adopting a strategy per se does not imply that possibilities of these kinds exist and, if so, concretely what they are; such matters can only be settled at the second moment where impartiality is the reigning aspiration.

Thus, neutrality cannot generally be counted on to hold; on application, at the third moment, theories will tend to serve especially well the sv linked with the strategy under which they are accepted. Nevertheless, I think that neutrality should remain an aspiration in the institutions of science, but now understood as: AS should be pursued in such a way that scientific knowledge is produced so that projects valued in the light of any viable sv can be informed, more or less evenhandedly, by well-established scientific knowledge. Generally, that neutrality is actually lacking will be testimony to the fact that AS has been pursued largely in response to the first pull that I identified in section 3.3. The thoroughgoing pursuit of AS requires the adoption of a plurality of strategies. I doubt that this can happen without the role of sv at the first moment being recognized as legitimate, and strategies linked with the sv of less dominant groups being provided with appropriate material and social conditions for their development (section 10.1; section 10.4).

Roles for Values at the Different Moments

Before concluding this chapter, it may be helpful to elaborate some points pertinent to the role of sv at the different moments. In the first place, although OSV, regardless of the sv that may be considered, does not play a logical role alongside the cv, nevertheless at the second moment sv may play various roles. These include: (1) Institutions that manifest certain sv may have cognitive value; (contingently, temporarily) they may provide conditions necessary for accepting theories in accordance with impartiality. (2) sv may be part of the causal explanation of why accepted theories of certain domains of phenomena are available, but not of others. (3) Adequate testing of theories—and especially the specification of the limits of the domains of phenomena of which they are soundly accepted—may require critical comparison with theories developed under a competing strategy that has mutually reinforcing relations with particular sv. (4) Since a theory may be applicable so as to especially favor one value-outlook, or it may undercut the presuppositions of another, commitment to sv (that are not served by the theory's application or whose presuppositions are undercut) can lead to raising the "standards" for "measuring" the degree of manifestation of the cv (section 9.4; section 9.6). (5) Holding particular sv may attune us to diagnosing when a theory is being accepted in discordance with impartiality, such as when an OSV is in fact covertly playing a role at the second moment alongside the cv.[7]

Second, accept/reject is not the only stance relevantly taken towards T in scientific activity. T may be provisionally entertained, pursued with a view towards its development or revision, subjected to testing, held to be more promising or to "save the phenomena" better than extant alternatives, used instrumentally in other inquiries, etc. Clearly some of these stances must have been adopted at earlier stages of the research processes that produce a soundly accepted theory. (Some theories are never candidates for acceptance—"ideal" theories, some mathematical "models.") The model of scientific activity proposed here may be elaborated to include further moments and submoments to correspond with these stances. At some of these, sv may have proper roles. Once a strategy has been adopted at the first moment, for example, there is a moment at which specific problems for investigation are chosen. Even those who affirm autonomy (and who do not recognize that there is a question of choice of strategy) readily admit a role for sv at this moment.

Third, I have maintained that application is an important moment of scientific activity, so much so that the sc served by application may also play a role at the moment at which a strategy is adopted. Thus, sv play a role at the core of scientific activity, and I see no good reason to want to eliminate them from this role. At the third moment, sv play a variety of roles connected with the legitimacy of applications. Legitimacy requires attention to judgments about NSE and NBW that are not reducible to those of theory acceptance. I will say that scientists "endorse" (or not endorse) hypotheses of the types NSE and NBW. To endorse a theory or hypothesis involves appeal to both cv and sv. Those who want to keep sv out of the heart of scientific activity do not consider judgments of endorsing to be proper scientific judgments. This is a pretty implausible claim (Machamer and Douglas 1999; Douglas 2000). Scientists, qua participants in scientific practices, do and are expected to tackle problems connected with NSE and NBW.

To endorse T is to judge that T has sufficient cognitive value (i.e., it is sufficiently likely to be true, or to become soundly accepted) that the possibility of future research leading to its rejection (or, its being false), and the possible consequences (serious negative ones from the perspective of specified sv) of acting on it if this were to happen, should not be considered good reasons not to engage in actions informed by T (section 9.6; cf. Rudner 1953). Endorsing T is a necessary (but not a sufficient) condition for the legitimacy of its application or of acting informed by it. Acceptance implies endorsement, but we cannot reasonably expect acceptance as a general condition on endorsement: we have to act in the absence of not only certainty, but also of knowledge meeting the high standards needed for sound acceptance. This is particularly relevant where matters like NSE and NBW are pertinent. When we endorse without acceptance, sv are always in play whether recognized consciously or not, and scientists' judgments may differ because of the different sv they hold. Endorsement is a moment of scientific activity (a submoment of the third moment, which also has impact on what strategies should be adopted at the first). To deny endorsement to T, when other scientists endorse it, carries the obligation to specify what further testing is needed (chapter 9; chapter 10; especially section 9.4). If none can be specified after a due lapse of time, then T has been tested according to the highest and most rigorous available standards. Science can do no more.[8]

Not Keeping Values Out of Science

I am indeed interested that the theoretical products of science have sound (reputable) cognitive credentials. I share this interest with the traditional view referred to at the beginning of this chapter (see also section 5.1). But I do not want to keep values out of science. Values are already there. My recommendations are for reform. I want to let in more values (section 4)—at the proper moments, and thus to dull the influence of the values that are already there. No doubt, my recommendations may allow some unsavory characters to get their foot inside the door (posing as my friends), but the model I propose of scientific activity gives plenty of resources to slam the door hard on them. This model enables us to understand certain phenomena of current scientific activity and to underwrite recommendations about how to improve it. It requires the distinction between cv and sv, and reflects simultaneously cognitive and political interests.

NOTES

1. In *SVF* chapter 3 I introduce a more general notion of "cognitive value" that applies to beliefs held in ordinary life and that informs actions in a variety of practices. The narrower notion is sufficient for the purposes of this book. Then, "cognitive value" may be considered an abbreviation for "value of accepted theory" (ø = accepted theory).

2. There is no settled terminology to mark this distinction; see the references in *SVF*: 262.

3. Stating the aim as I have in AS is, of course, contentious. Others prefer, for example, to state it in terms of "problem solving" (Laudan 1977). I will not defend my formulation here (see *SVF*: chapter 5), and will simply note that I think that the bulk of the argument that follows can be rearticulated in the context of "problem solving" versions of the aims of science.

4. Another question may be raised here. Let me put it in quasi-paradoxical form: Even granted that cv are distinct from other kinds of values, is the same set of cv appropriately deployed regardless of the domain of inquiry (physical or social) or of the strategy adopted in research? Or, could it be argued that the cv are in some sense relative to the strategy adopted? This question, I believe, needs further exploration. The answer to it could have impact on my current views.

5. In section 1.1, I listed a number of presuppositions of V_{MC}. The evidence that is claimed for them draws in part upon the high degree of embodiment of V_{MC} in predominant contemporary institutions. Adherents of V_{MC}, drawing implications from these presuppositions, tend to consider the values of contesting movements to represent either abhorrent social visions or unattainable ideals. Contestants, taking issue with the evidential support of the presuppositions, trade comparable charges (chapter 11).

6. See chapter 9. Note that NSE and NBW obtain concrete instantiations in P_3 and P_4 (respectively) throughout part II.

7. Longino (1990, 2002) deploys items like (1)–(5) in order to raise the question of whether there is a significant distinction of the cognitive and the social. I cannot engage with her arguments here; see *SVF*: chapter 9.

8. The category "value appealed to in offering solutions to problems addressed by scientists" does not provide a ground for the distinction between cv and sv. Even so, the distinction between cv and sv is important in the context of appraising NSE and NBW. It underlies how disagreements about endorsement can be rationally addressed, and helps to explain why often they are not (chapter 9 and chapter 10).

Chapter Four

Incommensurability and "Multicultural Science"

In the previous chapters I defended a version of methodological pluralism, urging that scientific practices should be conducted under a plurality of strategies. In this chapter I will explore how wide the scope of this pluralism might be by discussing how, consistent with maintaining commitment to impartiality, cultural values may properly have impact on scientific practices. This will provide a fruitful context for raising the controversial issue of "multicultural science." In the scientific mainstream, it is often contended that this phrase secretes a contradiction. Against this, I will show how the account of the interplay of science and values proposed in this book provides for the possibility of multicultural science, that is, of legitimate culture-based variations in approaches to scientific practice. Proponents of multicultural science often draw upon Thomas Kuhn's views about the incommensurability of theories developed within incompatible paradigms. I will follow this lead. Doing so will also enable us to gain a richer sense both of the ways in which different strategies may be seen to compete, and of how they may complement one another.

4.1 COMPETING STRATEGIES

In section 1.3, I likened the idea of "strategy" to Kuhn's "paradigm." A paradigm includes a strategy; and strategies include Kuhn's later innovation, "structured lexicons" (Kuhn 2000). Strategies provide the relevant features of paradigms that enable competing ones to be incommensurable. Recall that adopting a strategy is inseparable from delimiting the realm of phenomena that is considered of interest for investigation and the kinds of possibilities that are desired to be encapsulated; and so the reasons for its adoption may include metaphysical and value commitments. Under materialist strategies, we saw, theories are constrained to represent phenomena and possibilities as being generated from underlying structure, process, interaction, and law, so as to encapsulate the decontextualized possibilities of things. Aristotelian strategies, in contrast, involve relating phenomena to their places in the cosmos, and their theoretical categories reflect organic metaphors while relevant data may deploy, for example, common sensory and teleological categories. Agroecological strategies (the key to my discussion of incommensurability and multicultural science) are also different. (They are characterized in detail in section 5.4.) They do not abstract from the social, human, and ecological context and dimensions of things. Their focus is upon productive and sustainable agroecosystems and their constituents (seeds, plants, etc.) and "complex interactions among and between people, crops, soil, and livestock" (Altieri 1987: xv), whose possibilities cannot be reduced to decontextualized ones (at least to those that can be encapsulated under currently available theories).

A major theme of the previous chapters has been that modern scientific inquiry has been conducted virtually exclusively under materialist strategies.[1] Indeed, I have already pointed out that probably there are no strategies entertained today, including those implicated in my argument below for the possibility of multicultural science, that would not allow within them an important role, albeit a subordinate or circumscribed one, to some uses of materialist strategies. In order to highlight this fact, it might be a good idea to supplement my earlier characterization of "science"—systematic empirical inquiry, conducted under any strategies, that holds its results to accordance with impartiality—with the additional qualification, that it be inquiry that in principle (where appropriate) can draw upon knowledge gained under materialist strategies.

Adopting a strategy is necessary for the conduct of research, and research conducted under different strategies (provided that it is fruitful, i.e., successful in constructing and consolidating theories that manifest the cognitive values highly) enables us to identify different classes of possibilities. Under the strategies of agroecology, for example, are identified the possibilities of producing crops, so that people in the region of the production will gain access to a well-balanced diet in a context that enhances local agency and well-being, and that sustains the environment; under those of biotechnology (instances of materialist strategies), the possibilities of maximizing crop production under conditions— use of fertilizers, pest and weed management techniques, water, machinery, strains of seeds, etc.—that can be widely replicated. In general the possibilities encapsulated by theories developed under different strategies at most overlap, suggesting that a "complete" understanding of the world cannot be obtained (even in principle) under one kind of strategies. This is the ground for the complementarity of strategies.

Strategies and the "Worlds" in Which They Structure Research

Strategies also compete; conducting scientific research under one strategy may be incompatible with conducting it (in the same social context) under another. (Kuhn says that each paradigm defines its own "world."[2] I prefer the metaphor of "playing field.") The prescriptions (concerning constraints on theories and selection of data) of different strategies are practically incompatible; they structure "incompatible modes of [scientific] community life" (Kuhn 1970: 94). They cannot be followed simultaneously within the same practice. The application of the structured lexicon of one strategy (e.g., one that deploys the categories of quantity and law) may preclude that of another (e.g., one that deploys sensory and teleological categories). In addition, interactions with natural objects needed to obtain prescribed kinds of data for one may interfere with the conditions needed for obtaining the kinds of data prescribed by the other. For example, the interactions involved in investigating crop production as a function of widely replicable methods may interfere with maintaining local ecological stability; and engaging in experiment involves modification of "natural" settings, the observation of which provides prescribed data under some strategies.

I take it that what Kuhn called "incommensurability" (Kuhn 1970: 150; cf. Hoyningen-Huene 1993: 208–12) may be identified with this practical incompatibility of strategies. It is also related to claims about the "incommensurability" of theories. We do not refer to theories, such as those of quantum mechanics and evolutionary biology, which clearly are about different things, as incommensurable (Hoyningen-Huene 1993: 218–21; Carrier 2001). And do we say that the strategies under which they became accepted are incommensurable? Some strategies are simply different; others compete—their "playing fields" impinge on one another, threatening to undermine the "games" being played; their respective scientific "worlds" clash in the shared social "world" that nourishes them (*SVF*: chapter 7). Only where strategies compete do we speak of "incommensurability."

Ways in Which Strategies May Compete

There are several (not mutually exclusive) ways in which strategies may compete. In the first place, concerning application to phenomena of our shared social world:[3] Numerous interesting phenomena are described using the lexicons of common idiom that are shared by most people regardless of their theoretical orientations. In order to explain these phenomena recourse is had to available scientific theories. Theories are applied to them for explanatory and predictive ends; in doing so, objects described in common idiom are identified with objects articulated theoretically. Contradictions may be generated (in the lexicon of common idiom when it incorporates the theoretical lexicons) from attempts to apply theories developed under different strategies to the same phenomena of our common experience (*SVF*: 161–67).[4]

Second, concerning application in practical projects: Theories produced under different strategies may encapsulate possibilities of the same objects (e.g., seeds)—for example, those of agroecology and those of biotechnology—that cannot be co-realized (to any significant extent) in our shared social world; on application they may inform human practices in fundamentally different ways serving incompatible value-outlooks (part II). Then there is competition about which strategy's knowledge will be applied. In turn, since research is dependent upon the availability of relevant material and social conditions, there will also be competition for the resources and conditions needed for research under the strategies. The outcome of this competition is strongly influenced by the social values and forces with interests in the applications. Social competition almost inevitably leads to certain strategies remaining underdeveloped or to competitors to predominant strategies not being entertained (section 2.1).

Third, with roots in conflicting worldviews and value-outlooks: Far-reaching assumptions (about the general character and possibilities of things) that are drawn upon to support adoption of strategies, or to legitimate applications of their theoretical products in practical projects, may be inconsistent. (a) Conflicting metaphysical assumptions, for example, were involved in seventeenth-century competition between materialist and Aristotelian strategies. (b) More immediately, there are assumptions that reflect fundamental value commitments of dominant contemporary social forces and institutions (rather than the outcomes of empirical inquiry), and that have become deeply ingrained in the "common sense" of our times, that serve both to legitimate the expanded role of biotechnology in agriculture and to support adoption of biotechnological rather than agroecological strategies in research (part II). Among them are assumptions (including the presuppositions of the modern valuation of control) like "the inevitability of economic globalization" (chapter 11) and (more specific versions of these presuppositions) like "feeding the world's rapidly expanding population requires the development and implementation of biotechnology-informed agriculture." Assumptions of these kinds become presuppositions of value-outlooks that are dialectically related with the strategies adopted in research.

Competition of the second and third (part b) types has its origin in the third moment of scientific activity (section 3.4), that of applying scientific knowledge, and reflects that values in play at this moment may feed back and influence the first moment, choice of strategy. Kuhn considers only competition of the first and third (part a) types. This reflects his views that competition among strategies occurs (or should occur) only at times of "scientific revolution," and that social factors, including those relating to applied science—though necessary for investigation to proceed and perhaps to motivate the initial exploration of certain strategies—are not among the grounds for adoption of the strategies that guide "normal science" (section 4.3; section 5.2).

Kinds of Incommensurability

The practical incompatibility of competing strategies includes "semantic incommensurability" of theories as one of its features: Posits in theories developed with the structured lexicons of different strategies, and their negations, may bear no entailment relations with one another, since the lexicons may not be isomorphic and thus relevant items of the lexicons cannot be intertranslated. How widespread the phenomenon of semantic incommensurability is, where key theoretical terms from different lexicons actually have different meanings, is a matter of dispute. Kuhn himself narrowed the scope of his claims about it considerably in writings after *The Structure of Scientific Revolutions* (Kuhn 1970).[5]

The practical incompatibility of competing strategies is also the source of "methodological incommensurability," which I identify as having two components. First, the degrees of manifestation of the cognitive values in theories (T_1, T_2, etc.) developed under competing strategies cannot (generally) be compared, so that theoretical conflict across strategies cannot be rationally resolved according to the model, "Where T_1 and T_2 conflict, T_1 is rationally preferable to T_2 if and only if it manifests all (most?) of the cognitive values more highly than T_2 does."[6] Second, choice among competing strategies is underdetermined by considerations that rest solely upon the cognitive values. I think this explicates what is sound in the notion of "methodological incommensurability." Others have proposed to explicate it more or less as follows: The methodological standards for appraising theories are different within different paradigms—where, under "standards," cognitive values and strategic prescriptions tend to be grouped together indiscriminately. Often this indiscriminate grouping occurs as part of an argument that the adoption of methodological standards, including cognitive values, reflects social and cultural contingencies (e.g., Doppelt 2001). This is a mistake. Cognitive values (allowing for some historical development) may and ought to be shared across strategies, so that adopting them is not open to the same kind of social and cultural explanation that can be appropriate for strategic prescriptions (but, see chapter 3: note 4). Adopting incompatible strategies is consistent with holding that theories developed under all strategies should be appraised for acceptability (though not necessarily for being provisionally entertained or being considered worthy of further investigation or of application) in the light of the same set of cognitive values. It is important to separate the (logical) moments of adoption of strategy where social and cultural values have legitimate play, and of theory acceptance, which rests only on the data and the cognitive values (section 3.4). This permits that theories developed under strategies that compete, especially in the second way, may encapsulate different classes of possibilities of the same things, so that strategic competition need not always produce theoretical competition.

4.2 COMPARING THEORIES CONSTRUCTED UNDER COMPETING STRATEGIES

In order to see why the model of theory choice mentioned in the previous paragraph is unable to resolve theoretical conflicts that cut across strategies, consider the cognitive value, empirical adequacy. It gains a concrete interpretation, which enables the degree of its manifestation in a theory to be "measured," only in the context of a strategy that prescribes what kinds of data should be brought into contact with what kinds of theories. A theory is not simply empirically adequate but empirically adequate with respect to specified kinds of data. But the specified kinds vary with strategy. Thus, across strategies, we cannot (generally) compare empirical adequacy, for the only relevant comparison occurs when the data that one theory (should) fit constitute a subset of those that the other (should) fit. Similar conclusions can be drawn about the other cognitive values that concern relations between theory and data. Thus, the model applies (generally) only to the comparison of theories developed under the same strategy.

Another model can apply to competing fundamental theories T_1 and T_2, generated respectively under competing strategies S_1 and S_2: T_1 is rationally chosen if and only if T_1 manifests highly the cognitive values interpreted according to the (constraint/selection) prescriptions of S_1, and T_2 manifests (has come to manifest) many of them to a low degree interpreted according to the prescriptions of S_2; and S_2 has ceased to be very fruitful, in other words, efforts under S_2—supported by appropriate material and social conditions—to generate theories that manifest the cognitive values highly are not (have ceased to be) successful (*SVF*: 229). Where this second model applies, accepting T_1 is inseparable from discarding S_2 as a strategy rationally worthy of adoption, thus from ruling out hope of finding a "better" successor to T_2 through further research under S_2. For Kuhn, a theory choice made according to the second model is rational; and it plays a central role in his account of the rationality of "revolutionary" transitions.[7] There is, however, no assurance that a comparison of theories developed under competing strategies can always be made according to it—if, for example, both strategies are fruitful to some extent, as Kuhn suggests tends to be the case during a "revolutionary" period.

In the light of both models, accepting a theory is portrayed as choosing it rather than a competitor. In the light of the second one, it is also implicated in discarding the strategy under which the latter has developed; but (contrary to Kuhn—see next section) not necessarily in adopting (for ongoing research) the strategy under which the former has developed. Adopting a strategy is inseparable from identifying the kinds of possibilities that are desired to be encapsulated in theories. When a theory is soundly accepted, that is, when it manifests the cognitive values to a high degree of certain domains of phenomena, then, consistent with either model, it successfully encapsulates relevant possibilities of these domains. This remains untouched, even if the theory later is displaced by another, which manifests some of the cognitive values more highly.[8] Sound judgments made according to both models accord with impartiality; their grounds are independent (logically) of cultural, ethical, and social values. Making such judgments, however, may depend (causally) upon adopting particular social values, for to make them one must have developed the necessary cognitive abilities, be suitably located, or have an interest in doing research on which they would be based. It is important to distinguish the grounds upon which such judgments are rationally based and the factors that explain that the conditions are available for them to be made (*SVF*: 231–36). The latter, but not the former, may include social and cultural values. Note also that, in view of the second type of competition referred to above, there can be serious practical impediments to using the second model to make impartial judgments (section 3.4).

Nevertheless, it does not follow, where the second model is operative, that the strategy under which the chosen theory has developed should be adopted for the conduct of ongoing research by every investigator regardless of their social and other values. For there may be yet another strategy (S_3), which remains fruitful (or would become fruitful if research conducted under it were supported with appropriate material and social conditions), and which leads (or would lead) to understanding deemed more significant (of more social value), that is, more applicable to valued objects and in valued social activities. Accepting a theory according to either of the two models does not imply that it has significance independently of the cultural, ethical, and social values held (section 1.2).

4.3 CHOOSING TO ADOPT A STRATEGY

When adopting a strategy it is (rationally) proper to consider both fruitfulness and significance. For Kuhn, fruitfulness is sufficient, since he maintains that research (in a given field) normally is conducted under a single strategy, adopted in the aftermath of theory choices made according to the second model, accompanying which a nonfruitful strategy is discarded, leaving one fruitful strategy in place to be pursued until its fruitfulness is exhausted. It is indeed true that fruitfulness suffices to rationalize the adoption of materialist rather than Aristotelian strategies, but not necessarily rather than any other strategies (*SVF*: chapter 7). Perhaps at the time of "the scientific revolution" the pertinence of significance was not apparent (though it seems to have been for Bacon with his emphasis on "utility"), for then—in the light of the emerging values of modernity—materialist strategies were widely deemed highly significant, as they still are. Then perhaps the issue of possible alternative strategies, other than the discarded Aristotelian ones, simply did not arise; and this may appear to be confirmed by the fact that the theoretical products of research conducted under materialist strategies are applied in the projects of leading social forces and apply to the technological objects that have come to figure so centrally in our lives. In general, theories developed under materialist strategies are highly significant wherever modern techno-scientific and economic values, in particular the modern valuation of control, are widely held. In modernity, then, both fruitfulness and significance point towards adopting materialist strategies (*SVF*: chapters 6 and 7). But they can be pried apart.

The modern valuation of control and the associated values of modernity (unless highly qualified) are not held universally—for some, because of the social and ecological devastation that modernization is perceived to have borne in its wake (chapter 11); for others because these values contribute to undermine "traditional" values and important dimensions of human well-being that might provide the basis for alternative forms of "development" (section 6.3; chapter 11; *SVF*: chapter 8; Feyerabend 1999: part II, chapter 9). For those who question the modern valuation of control, the general significance of understanding gained under materialist strategies is compromised, though the fruitfulness of materialist strategies remains established and alternative strategies may freely draw upon the positive knowledge gained under them. That is a good reason for them to seek out alternative strategies that may be able to generate acceptable theories that are more significant in light of their values, theories that might be applicable in their preferred ways of life and to important phenomena and possibilities in them.[9] The success of their quest is not guaranteed, but neither is its failure. There may be two fruitful strategies that enable the identification of possibilities of the same objects, but different classes of them, such as their possibilities as relevant to technological control, and those connected with furthering the manifestation of competing social values (to which technological implementations are considered subordinate—chapter 10). Whether there are or are not such fruitful strategies can only be known after appropriate research has been conducted under them.

Same Object Investigated Under Different Strategies

As pointed out in section 1.3, many objects, including experimental and technological phenomena, whose decontextualized possibilities are well grasped under materialist strategies, are also social objects, objects of social value. In important cases certain decontextualized possibilities cannot be actualized (in historical context) without also actualizing particular social possibilities and undermining others; and, in these cases, both kinds of possibilities can (in principle) be investigated systematically and empirically. For example, the possibilities of transgenic seeds, identified by the recent biotechnological research, cannot be actualized (under current conditions) without furthering the social process of turning seeds into commodities, and this not only changes the character of farming but also has profound ecological and social implications. That certain kinds of possibilities of transgenic seeds have been identified has been established in accordance with impartiality; that seeds become turned more completely into commodities also has (section 7.2; section 9.5). Both are soundly accepted claims established in the light of available data and the cognitive values; they cannot be properly challenged on the basis of the social values one holds. The commodification of seeds, however, does not represent a universal value, and so ceteris paribus the research practices that increase our grasp of the possibilities of transgenics are not universally considered of high social value.

Consider the two kinds of possibilities (described in section 4.1) about crop production, the first about it serving local well-being and sustaining the environment, the second about maximizing it under widely replicable conditions. Those who adopt materialist strategies (including those of biotechnology) exclusively are effectively able to address questions about the second, but questions about the first cannot be addressed if one abstracts from the social and ecological contexts. Questions about the first, however, will have greater salience to those whose values are in conflict with the modern valuation of control and the socioeconomic values along with which it is usually embodied in powerful modern institutions (section 6.3). Thus, since the questions are open to empirical investigation under agroecological strategies, these strategies will rightly take precedence for them; rationality does not require that they wait until the fruitfulness of the competing kinds of materialist strategies has been exhausted. Competing strategies, if they were to gain the material and social conditions necessary to develop, might all be fruitful; they might encapsulate different classes of possibilities of the same objects (e.g., seeds).

A Model for Deciding Which Strategy to Adopt

Moving beyond Kuhn, I propose a third model, one motivated by item (iv) of my statement of the aim of science, AS, in section 3.2: Only adopt a (potentially) fruitful strategy; but if competing strategies S_1 and S_2 are both (potentially) fruitful, adopt the one that may produce understanding of significance for the cultural, ethical, and social values that one rationally holds. In order to safeguard that this model will be applied consistently with the pursuit of objectivity (section 2.1), applying it would require antecedent appraising (and subsequent ongoing monitoring), in the light of relevant empirical investigation, of the likely (and potential) impact on human well-being of conducting (and providing material and social conditions for) research under it and of its potential applications. This model is proposed as one for an individual investigator (or group of investigators) to follow. I continue to maintain my earlier advocacy of a plurality of value-outlooks being represented within the worldwide community of investigators. Then, it is clear that the model is compatible with the strategic pluralism that I have defended. In practice, I think that this model should be deployed within a loosely knit body of institutions where there is sufficient oversight to ensure that an adequate plurality is supported. Working this out is one of the issues in establishing a proper balance between scientific autonomy and democracy (section 9.5; section 10.4).[10]

The third model, of course, is not a model for theory choice. Its use does not support judgments that theories that might be developed under a discarded (non-adopted) strategy are false, or that theories that fit the constraints of the favored strategy are acceptable. Neither does the use of this model lead to the definitive rejection of one of two competing strategies; and investigators, who hold different value-outlooks, may use it to support adopting different strategies—then the dispute between them is located at the level of moral, ethical, and cultural values and not in cognitive value judgment.

Difficulties in Deploying the Model of Strategy Choice

Making use of the third model confronts serious difficulties in practice (cf. section 2.1). I listed (section 4.2) three ways in which strategies may compete. When they compete concerning application in practical projects, the accompanying competition for resources may leave one of the strategies significantly underdeveloped (of interest only to marginalized groups), so much so that its potential for fruitfulness may not be discernible. Then, one strategy may be followed as if it has no legitimate competitors, thus disguising the role that values play in supporting its adoption. This situation is reinforced when the competing strategies also draw upon conflicting views, one set of which consists of presuppositions of the value commitments of dominant social forces and institutions.

The combination of the two ways of competing may render proposals developed under strategies that compete with the dominant strategies virtually unintelligible. Think of those forms of agroecology that maintain explicit continuity with traditional agricultural practices in some of the poor regions of the world (section 5.4; part II). On the one hand, if agroecological research gained the material and social conditions needed to explore fully and systematically the possibilities of crop production serving local well-being, and it were successful (section 10.1), that would threaten the increasing dominance of agribusiness in agriculture (section 10.3) and challenge some of the presuppositions of "globalization" (section 11.4), all of which are well served by scientific developments (e.g., in biotechnology) made under materialist strategies. On the other hand, the categories deployed with materialist strategies have gained such a grip on contemporary imaginative and conceptual powers that the idea that there might be serious alternatives to research under materialist strategies, under which soundly accepted theories might be consolidated, simply tends not to be considered (*SVF*: 126–30). Then, in effect, the prescriptions of materialist strategies become treated as if they were cognitive values.

The tendency to mistake dominant strategic prescriptions for cognitive values partly accounts for the lack of recognition of the role that the third model should play (chapter 2). This mistake may be reinforced by the fact that rarely is a strategy adopted as a result of explicit deliberation, but rather (in the first instance) generally an investigator enters into research activities under established strategies, so that a novice learns how to follow strategic prescriptions and how to estimate the degrees of manifestation of the cognitive values at one and the same time. Thus we should be wary of defenses of dominant strategies of the "It's the only game in town" type. It may be the only game in town because (whether deliberately or without awareness) competitors have been suppressed, not recognized, or denied resources for their development (section 7.2). The reasons that support adopting a strategy generally are articulated post hoc; then the role of the model may become apparent, for empirical inquiry under any one type of strategies is clearly inadequate to establish directly that all possibilities are of the kind that can be encapsulated under these strategies.

Grounding the adoption of a strategy with arguments that deploy the third model explicitly is likely to lead to the recognition that under the strategy various classes of possibilities will be left unexplored, and that assumptions about some of the unexplored possibilities are among those that support both the value of adopting the strategy and the legitimacy of applying its products. For example, the application of transgenic technology—and indeed adopting the strategies of biotechnology research itself—is often legitimated by appeal to the assumption "It's necessary to feed the world's rapidly growing population" (chapter 10), which in turn assumes that other agricultural approaches, including those of agroecology, cannot produce the required food (section 10.2). These assumptions are not the outcome of research conducted under materialist strategies, and cannot be; they could only be established with research that involved the significant development of strategies such as those of agroecology (section 10.4). It is not enough to show that transgenic technology provides a range of benefits (section 8.1; section 8.2), for that is compatible with agroecology being a comparable or superior competitor and also with the harmful side effects of applying transgenic technology outweighing its benefits.

Neutrality and Methodological Pluralism

In the light of the conclusions of the last two subsections, the interest of increasing the manifestation of neutrality in scientific practices would give rise to a community of investigators which upholds the methodological legitimacy and the possible fruitfulness of a plurality of strategies and provides (to the extent possible) conditions for the development of a wide array of strategies (including agroecological as well as biotechnological ones). It should also lead to regarding disputes about competing strategies as a normal part of scientific activity, and to expecting and permitting applications to be informed fully by knowledge gained under various strategies. It should engender tolerance to provide as much social space as possible to explore alternative strategies, and thus encourage investigators to temper commitments they have to the values, which are linked to the strategies they adopt, with humility and tolerance for other approaches (both scientific and socioeconomic). This is the context for the proper use of the third model. (I do not suggest that this context can be brought about simply by choices and judgments made within the scientific community without broader changes in socioeconomic ordering.)

No paradoxical kind of relativism is thereby implied. It is just that under the different strategies (in principle) different classes of possibilities are encapsulated that may not be able to be co-actualized in the same places and projects. Furthermore, one may uphold pluralism of strategies in the scientific community while choosing to adopt particular ones oneself; and do so while recognizing that research conducted under materialist strategies is capable of indefinite expansion, in part because it is of the nature of experiment and technology to create phenomena and the spaces in which certain phenomena are encountered, understanding of which is produced under these strategies. But indefinite expansion of material possibilities is compatible with not all possibilities reducing to decontextualized ones. I foresee, for example, no limit to the realm of possibilities that might be uncovered by research in biotechnology, but that provides no reason to hold that the possibilities open to things in virtue of their locations in agroecological systems are reducible to the decontextualized possibilities of the constituents of these systems (section 5.4; section 7.2; section 10.3).

4.4 "MULTICULTURAL SCIENCE"

Thus the door is opened to "multicultural science": Values derived from different cultures and embodied in radically different forms of life point to the salience of questions, open to empirical address, about material objects (e.g., seeds) that are not dissociated from their place in human experience and social structures, which are sidelined by the mainstream of modern science. They thus point to the potential value of identifying alternative strategies, which may involve rich developments of the strategies deployed in gaining "traditional" knowledge" (section 5.4; section 7.2). In principle, there is no reason why there should not be fruitful strategies that are incompatible with uniquely privileged materialist strategies and that compete with them concerning application in practical projects. Furthermore, concern for social justice, informed by the values of various cultures, can lead to anticipating the potential significance of products such strategies might generate; for example, they might successfully identify novel possibilities about crop production and local well-being (section 5.4; section 10.1).

I do not know how far the claims of multicultural science can be pushed. That needs to be established case by case in the light of careful investigation. While I have no doubt that some enthusiasts have exaggerated them, prima facie that a culture's cosmovision includes a view like "that nature is benevolent" (an example given in Siegel 2001) is not per se a reason to dismiss the products of the systematic empirical (scientific) inquiry associated with it—unless the view is shown to be inconsistent with a theory soundly accepted under materialist (or other) strategies.

It is not enough that it is inconsistent with materialist metaphysics, which may be taken to be the view that the decontextualized possibilities of things exhaust their possibilities (section 3.3). This view has pervaded the modern scientific tradition, whose early developments were inseparable from the simultaneous and reinforcing developments of this metaphysics. Galileo and Descartes, for example, offered a priori arguments that the way the "natural world" is makes it open to being grasped (ideally) under versions of materialist strategies. Still today commitment to materialist metaphysics lies behind the widespread upholding of materialist strategies, particularly reductionist versions that dominate the philosophy of mind and the practices of "scientific" psychology (Mariconda and Lacey 2001). Its categories (continually refined as they are extrapolated from the latest scientific developments) seem to dominate the imaginations of contemporary scientifically minded intellectuals, as if they must be constraints on our investigations. Why this should be is unclear. Few subscribe any more to a priori arguments to support materialist metaphysics.[11] Furthermore, materialist metaphysics is not a presupposition of systematic empirical inquiry (since there may be fruitful strategies linked with presuppositions that are inconsistent with it); and, as an extrapolation from the results of inquiries conducted under materialist strategies, it cannot ground their privilege. My own hunch is that the grounds for subscribing to materialist metaphysics are the same as those for the virtual unanimous adoption of materialist strategies, viz. dialectical relations with the modern valuation of control (*SVF*: 126–30). Allusion to materialist metaphysics tends to underlie common arguments against the possibility of multicultural science. Harvey Siegel, for example, builds his argument on the claim that (all) scientific research has a common object (an ahistorical one—cf. section 5.1), "the natural world," or parts of it, or objects in it "that have the properties they do independently of the cultural contexts in which people study them, [where] such properties are best studied and understood in . . . the ways recommended and exemplified by 'Western' methodological principles and practices (i.e., under materialist strategies)" (Siegel 2001).

Kuhn, of course, disagrees with this. For him the objects investigated under different strategies are largely different, occupants of "different worlds," which are partly constituted by and in the research activities themselves. And I, not departing very far much from Kuhn, have suggested that the object of inquiry conducted under materialist strategies may be considered to be the decontextualized possibilities of things. In all cultures some of the decontextualized possibilities of things are salient (since no possibilities are realized without also realizing certain decontextualized possibilities), and that is why I expect overlap between materialist and any other strategies concerning what the object of inquiry is. Nevertheless, these possibilities do not exhaust the possibilities of things insofar as they make contact with our lives (though perhaps they do, in principle, of the possibilities of technological control); understanding the "natural world," the "world in which we live," does not reduce to understanding gained under materialist strategies. Furthermore, theories developed under different strategies may compete in their accounts of what possibilities are realizable in the "world" of daily life by and in interaction with the objects encountered in the actual social world.

In multicultural science—more exactly, under a range of strategies reflective of a variety of cultural outlooks—material objects may be investigated for the social (including agroecological) possibilities in which they are implicated. If science is systematic empirical inquiry that enables us to grasp "the world in which we live" and that serves to inform our practical activities, there is a certain urgency to exploring concretely the potential scope of multicultural science and no reason to accept that a unique strategy will pass the test of fruitfulness. Of course, given that strategies linked with different cultural values will compete with materialist strategies concerning application in practical projects, major barriers to such exploration can be expected to remain in place (section 10.3–section 10.6; chapter 11).

That there are practically incompatible competing strategies, all of which may be fruitful, is the key to my argument. They are the source of the methodological incommensurability (section 3.2). But fruitful competing strategies need not generate competing theories; their respective soundly accepted theories might encapsulate well-different classes of possibilities of the same phenomena, whose realization in practical applications may serve different cultural, ethical, and social values. Thence, competing strategies are also the source of the possibility of multicultural science. Reflecting on this common source of methodological incommensurability and the possibilities of multicultural science also enables us to discern the impediments to the latter's growth and even to the recognition of its intelligibility. I repeat that I do not know how far the claims of multicultural science can be pushed. But it does carry an unambiguous record of fruitfulness: The seeds developed in the course of traditional agroecology have become the sine qua non for the development of transgenic seeds (section 7.2 and references given there).

NOTES

1. Although this is closely linked with commitment to materialist metaphysics, that all possibilities are decontextualized possibilities (section 3.3; section 5.4), adopting materialist strategies to investigate particular classes of phenomena and possibilities does not imply that one is committed to materialist metaphysics (McMullin 1999; *SVF*: chapter 6).

2. The interpretation of Kuhn's use of the metaphor of "different worlds" has been much disputed. I, more concerned with explicating a Kuhn-inspired view that I regard as sound than with the details of Kuhnian scholarship, emphasize that a scientific "world" is linked with a "form of life," its required skills (habits, expectations, and sense of what is possible), its organizing structures, and its ways of actively engaging in research (cf. Rouse 1987); and that different "worlds" are all located in the shared sociohistorical "world," whose phenomena they may enable us to understand and whose practices they may inform (*SVF*: 149–54). Hoyningen-Huene (1993), on the other hand, offers a neo-Kantian interpretation, identifying scientific "worlds" with "phenomenal worlds" and "the world" with the epistemically inaccessible "world-in-itself."

3. For the distinction between "apply *to* phenomena in our shared social world" and "apply *in* practical projects," see section 5.3.

4. Cf. Hoyningen-Huene (1993): 221, on Kuhn's recognition that incommensurable theories may be compared in their "empirical potentials."

5. See Sankey (1997) and Hoyningen-Huene (1993) on the developments in Kuhn's views over the years; also Kuhn (2000).

6. For discussion of cross-paradigmatic (cross-strategy) theoretical conflict see *SVF*: chapters 7 and 9. Contrary to many interpretations of Kuhn, incommensurability does not entail relativism.

7. Cf. the analysis of Kuhn's account of the rationality of the theory choices that end a "revolutionary" period offered in Hoyningen-Huene (1993): 241–43. Carrier (2001) clarifies the logic of the second model; for further elaboration see *SVF*: chapter 7.

8. Cf.: "In so far as Newtonian theory was ever a truly scientific (i.e., soundly accepted) theory supported by the evidence, it still is. Only extravagant claims for the theory—claims that were never properly parts of science (i.e., were not about phenomena of which the theory was soundly accepted)—can have been shown by Einstein to be wrong. Purged of these merely human extravagances, Newton's theory has never been challenged and cannot be" (Kuhn 1970: 99). See also section 5.2.

9. Cf.: "Professionals dealing with the ecological, social, and medical parts of developmental aid have by now realized that the imposition of 'rational' or 'scientific' procedures, though occasionally beneficial (removal of some parasites and infectious diseases), can lead to serious material and spiritual problems. They did not abandon what they had learned in their universities, however; they combined this knowledge with local beliefs and customs and thereby established a much needed link with the problems of life that surround us everywhere" (Feyerabend 1993: xiv).

10. This model has clear affinities with several of Feyerabend's themes: the methodological importance of proliferation of theories (Feyerabend 1981); the multiplicity of scientific

traditions many of which involve interaction and continuity with "traditional" forms of knowledge; that such proliferation and multiplicity serve to gain access to possibilities that are important for "humanitarian" and "democratic" reasons, but which are otherwise inaccessible (Feyerabend 1993); that scientific theories are appropriately evaluated for their ethical value (significance) or their place in a way of life (Feyerabend 1999); and that the dominance of a single science (conducted under materialist strategies) threatens human well-being (Feyerabend 1993; 1999).

11. Taylor (1982) and, in a somewhat different way, Maxwell (1984), argue that the best explanation of the "success" of modern science is that it produces a superior understanding of "the material world," ("the natural world"). My detailed criticism is in Lacey (1998): chapter 4 and *SVF*: chapter 6. Compare Bunge (1981) for a very different perspective.

Chapter Five

The Social Location of Scientific Practices

I have maintained that adequate scientific methodology requires the development of a plurality of strategies, where adopting a strategy may have dialectical links with holding a particular value-outlook. This opens the possibility (chapter 4), building on arguments presented in the earlier chapters, that the range of strategies in play in scientific practices could be reflective of a variety of cultural outlooks. While my argument has mainly been presented in general terms, I have frequently illustrated it using the example of research conducted under agroecological strategies. The example plays a crucial role in my argument. (No doubt there are other examples from other fields that could play this role equally as well.) It serves to rebut the criticism that my argument suggests only an abstract possibility. It exemplifies that there really are alternative fruitful strategies, which have dialectical links with value-outlooks that contest the modern valuation of control, that are not reducible to materialist strategies. But, so far, the example remains underdeveloped, consisting of little more than occasional hints about what agroecology and its possibilities are.

In this chapter, I remedy this state of affairs, and offer a detailed account of agroecological strategies (section 5.4). Those who engage in agroecological research have a clear sense both of the non-reducibility of their strategies to materialist strategies, and of the dialectical links between adopting them and holding values like those of popular participation (section 6.3). Because of the explicit linkage with values, it has been not uncommon in the scientific mainstream to dismiss the "scientific" credentials of agroecological research. In order to counter this dismissal, I present my account of agroecological strategies in the context of its competing with biotechnological ones, which have dialectical links with the modern valuation of control. In the process, the methodological pluralism that I have defended is confirmed and provided with rich concrete content. It emerges from this (extending the argument of chapter 4) that it is difficult to separate the character of scientific research, that is, the strategies under which it is conducted and so the interpretation of its results, from the "social locations" in which it is conducted—from its relations with the value-outlooks of its practitioners and their enabling institutions and the interests that will be served through applications of their products. From this, it follows that scientific practices exhibit historicity: that their character changes, and must change, in fundamental ways that arise historically and that are responsive to and shaped significantly by historical and cultural variations in the realm of daily life and experience and in the structures of social practice. This chapter is structured around the question of the historicity of scientific practices.

5.1 SCIENTIFIC PRACTICES AS LACKING HISTORICITY

Let me briefly recapitulate a story that articulates an important part of the self-understanding of the modern scientific tradition, a story that has often been retold and that is not without considerable attraction. Science has a history, a history of progress: of growth, accumulation, and refinement of scientific knowledge, and of elimination of error. It is a history in which methodology plays a central role. Provided only that scientific practices are kept free from outside interference and nourished from time to time by the input of creative genius, methodology ensures the continued unfolding of the progressive development of science. Scientific methodology is systematic and empirical, rooted in experiment and measurement. It prescribes that empirical data be brought to bear upon theories that, using the resources of mathematically articulated lexicons, posit representations of phenomena and their underlying order and law. Apart from refinements of detail, scope, and precision, scientific methodology does not change. Thus, the "scientific revolution" of the sixteenth and seventeenth centuries marks the effective beginnings of science (anticipated only by scattered fragments of scientific knowledge), rather than a fundamental change in the methodological character of scientific practices.

Since then, the story continues, the cognitive (epistemic) credentials of scientific methodology have been certified and repeatedly vindicated. Technological success that has been informed by scientific knowledge has been one source of the vindication. Another has been the knowledge and understanding of "the world" ("the natural world," "the material world")—of natural laws, and of things, events, states of affairs, phenomena, structures and their underlying components, processes, and interactions—that have accumulated and been refined, and whose compass continues to expand, bounded only by the limit of a "complete account" of "the material world," one that in the limit would encompass all phenomena.

The story admits of competing versions with different emphases about, for example, the primacy of theory or experiment and the significance of applied science. In all versions, however, the tale of progress attends principally to such matters as theories that have been developed, available data, technical possibilities for experiment and measurement, methodological matters, and the (creative) inputs of individuals (or groups) of scientists. That way the "rationality," "universality," and "objectivity" of the cumulative and developing process are able to be emphasized. There is a place in the story for social, economic, and political factors: sometimes the interests of utility lead to a focus on a particular object of inquiry, and, more generally, the rhythms and organization of scientific research depend upon the availability of the appropriate material resources and social conditions. Even so, the fundamental dynamic of scientific progress is "rational": science is progressively, and with ever-greater refinement, gaining knowledge of objects of "the material world." Social (and hence historical) factors may lead to giving priority, even urgency, to gaining knowledge of specific instances of these objects, but gaining knowledge of them contributes to the overall accumulation and refinement of knowledge of "the material world." It is the accumulation and refinements that matter most. The rest, including the temporal order in which objects are investigated, is incidental. Nowhere in the story does the character of scientific practices change in fundamental ways. While they have a history, they do not exhibit historicity.

Assumptions Supporting the Denial of Historicity

At the root of this denial of historicity are the following three assumptions:

1. Science aims to gain a kind of understanding that is expressed in theories that match ever more completely and accurately an ahistorical object, "the material world," whose underlying order (laws; and structures and their components, processes, interactions) is ontologically independent of human actions, desires, conceptions, observations, and investigations.
2. The methodology of modern scientific practices (subject only to refinements of precision, scope, and the like, but not to any fundamental change) enables us progressively to gain understanding of this ahistorical object—so that there is no deep historical dialectic of methodology and object of inquiry, and so that the questions posed in basic science (while they might depend on the results of previous inquiries and the availability of instrumentation and appropriate mathematical and conceptual resources) do not concern objects insofar as they are historically variable, socially located, or playing integral roles in human practices.
3. The acceptability of scientific theories depends only on considerations involving their features and their relations with empirical data of selected kinds, as well as some intertheoretic relations.

Clearly, and consistent with Assumption 1, the actual arrangements of material objects in our vicinity are not causally independent of human affairs. These arrangements may be consequences of scientific applications; so much so that, although science supposedly lacks historicity, in virtue of its applications it has become nevertheless a historical agent of extraordinary importance. Indeed, it has been held, the very success of science in informing technological developments is explained in terms of its having gained sound understanding of "the material world" (see *SVF*: chapter 6). According to the story being recapitulated, the historical agency of science may account for the ready availability in the advanced industrial countries of the social conditions and material resources required for the pursuit of science. Moreover, nowadays a good deal of research depends upon the availability of instruments that are the products of the most advanced and sophisticated technology, whose availability is itself made possible by scientific developments. This means that the historical agency of science functions as an "instrumental partner" of scientific research, one that enables the methodology to be deployed in a more refined way (e.g., enabling us to obtain greater precision in measurement and to explore hitherto inaccessible spaces). It feeds back so as to serve the cognitive (epistemic) interests of gaining scientific understanding, a "happy coincidence" of social practical interests and knowledge-gaining interests (*SVF*: 124–26), but it leaves the fundamental character of scientific methodology essentially unscathed. Thus, the denial of historicity also involves the assumption:

4. The historical agency of science (exercised through its applications) is only a consequence and an instrumental partner of successful scientific practice; it is not a dialectical partner, one that feeds back so as to influence the fundamental methodological character of these practices.

Affirming the historicity of science involves denying Assumptions 1 and 2 denying that the object of scientific investigation is ahistorical and maintaining that there is a dialectic between methodology and object of inquiry. It is deepened by also denying Assumption 4, affirming that there is a dialectic between methodology and the practices of socially applied science. Thomas Kuhn has made a compelling case for the denial of Assumptions 1 and 2, though he seems to accept Assumption 4. (I do not take issue with Assumption 3; it is a version of impartiality.) Before addressing Kuhn's argument, let us extend the story being recapitulated a little further.

Do Soundly Accepted Theories Represent the Material World?

According to our story, objects as grasped in the practices of basic science, that is, objects as represented in soundly accepted theories, are (approximately) identical to objects as they are in the underlying order of the material world. Scientific practices, and the modes of interaction and thought that constitute them, enable us to grasp things as they are in the ahistorical "material world," abstracted from the context and conditions of our investigations, and indeed from all human related contexts. Since scientific practices are themselves historical, how can this be so? Methodology is the key to the answer. But how is it that a methodology deployed within a historical practice can enable us to grasp the ahistorical?

It is able to do so, the usual story goes, in virtue of the character of and relationships between theory and empirical data. Elaborating: theories developed in scientific practices deploy carefully expressed posits (and models), typically in mathematical form, about underlying (nonapparent) structure and its components, process, interaction, and law; so that theories dissociate (decontextualize) the phenomena investigated from their places in the social order, in daily life and experience, and even in scientific practices themselves. And data are sought out and reported, and the conditions in which they may be obtained are often created, in the course of experimental and measurement practices. Relevant data, obtained from observing phenomena of which a theory is proposed to provide understanding, meet the condition of intersubjectivity (and, where possible, replicability), and quantitative and experimental data are of special significance. Then understanding of objects of "the material world" is expressed in soundly accepted theories.

A theory is accepted if its posits (pertaining to certain domains of phenomena) are put into the stock of settled scientific knowledge, the stock of those posits judged to be such that further investigation or testing of them would produce at most refinements of accuracy and scope. A theory is soundly accepted (of the phenomena of a specified domain) if it satisfies certain criteria, viz. if it manifests the cognitive values highly in relation to the available data from this domain—if it has specified characteristics (e.g., consistency, simplicity), relations with other accepted theories (e.g., intertheoretic consistency, consilience), relations with displaced theories (e.g., being a source of interpretive power of the strengths and weaknesses of a displaced theory), and most importantly relations with available empirical data (e.g., empirical adequacy, explanatory and predictive power). (A theory is soundly accepted if and only if it is accepted in accordance with impartiality.) Theories which have been soundly accepted of specified domains have also reliably informed numerous practical (technological) applications.

What legitimates the move (where T is a theory and D is a domain of phenomena) made in the story from:

(a) T manifests the cognitive values highly with respect to D

to

(b) T represents (matches) order of the "material world" underlying D?

One might respond: Is it not obvious, given that T represents the phenomena of D, in abstraction from the relations they may have with human and social affairs, in terms of their being generated from the underlying order, and that the sound acceptance of T depends only on judgments of the manifestation of the cognitive values in T with respect to D? Moreover, that the move has been frequently and casually made throughout the course of modern science suggests that it is taken to be obvious. In what else could the cognitive value of T consist other than a match with parts of the "material world"? Well, it could consist in confirming

(c) T encapsulates well the possibilities of phenomena of D insofar as they may be represented as deriving from the generative power of the underlying order,

or (equivalently) in grasping these phenomena qua dissociated from their human and social contexts. In earlier chapters I called such possibilities the decontextualized possibilities of phenomena.

Elsewhere (*SVF*: chapter 6; Lacey 1998: chapter 1; 2002e), I have argued that the move from (a) to (b) is not soundly mediated by

(d) T reliably informs technological applications.

But the move from (a) to (c) is supported by (d). Often the move from (a) to (b) is made against background commitment to materialist metaphysics, which may be considered as a suitable elaboration of the posit that all phenomena are lawful or that all possibilities are decontextualized possibilities: the "material world"—the ahistorical order underlying things—really is such that it can be matched by (and only by) the kinds of posits put forward in modern scientific theories. Were there a sound a priori case for materialist metaphysics this might be compelling. But today, for the most part, those who espouse materialist metaphysics do so on the ground that it is an extrapolation from established scientific understanding and the direction of its expected growth. Then, if materialist metaphysics provides the ground for the move, the question is begged.

5.2 KUHN'S ACCOUNT OF THE HISTORICITY OF SCIENCE

Kuhn maintains that there is nothing in the character of scientific practices that justifies the move from (a) to (b) (*SVF*: chapter 7) and that, furthermore, attention to the actual history of science suggests that the move would be clearly unjustified. In the history of science, he maintains, we do not find steady accumulation and refinement, but instead, periods of fundamental discontinuity in the character of scientific activity—discontinuities, for example, in what is considered a theory worthy of provisional investigation, in what are the appropriate phenomena to investigate for the sake of gaining empirical data (and in the descriptive categories of the data) that are to be fitted by theories, and in what kinds of posits are taken to be central for shaping scientific investigation.

Soundly Accepted Theories: Developed and Consolidated under a Strategy

According to Kuhn, if theory and empirical data are taken to be the only major elements of scientific methodology, no sense can be made of the actual history of science. Kuhn proposed a third element: paradigm, of which I will consider just one aspect: within a paradigm, research is conducted under a strategy (section 4.1), which specifies constraints upon theories that are taken to be admissible for provisional consideration (and possible eventual acceptance), and (reciprocally) criteria for the kinds of empirical data (and the phenomena from the observation and measurement of which they are obtained) that are selected as those relevant for being brought into the appropriate relationships with theories. These are the kinds of data needed for testing and selecting among provisionally entertained theories, and those that describe phenomena so as to enable their explanations and the encapsulation of their possibilities. Admissible theories may be constrained, among other things, to be formulated with the resources of a specified lexicon, for example, the teleological/sensory categories of Aristotelian physics, or the mathematical/mechanical ones of Galilean physics; and the data may be selected (generally subject to the condition of intersubjectivity and, where appropriate, replicability) respectively (in the Aristotelian/Galilean case) in virtue of being representative of phenomena of daily life and experience, or of pertaining to experimental and measurement practices.

In earlier chapters I have pointed to two key (logically distinct) moments of choice: choice of strategy to adopt in research practices, choice of theory to accept or reject. Choice of theory is, then, in the first instance choice among provisionally entertained theories that fit the constraints of the adopted strategy.[1] When properly made, it involves judgment about which one of them best manifests the cognitive values with respect to the available data, about whether the available data are sufficient, and about whether the manifestation meets high enough standards for accepting the theory of the relevant domains of phenomena. Accepted theories encapsulate soundly certain kinds of possibilities that these phenomena permit. (Successful application testifies to this.) So, adopting a strategy involves identifying the kinds of possibilities desired to be encapsulated; accepting a theory involves identifying (typically through consolidating posits about how to actualize them) the genuine possibilities of these kinds.

In the light of the Kuhnian insight, our initial story can be reinterpreted or (more accurately) replaced by a narrative of research conducted under a particular set of strategies—materialist strategies (symbolized as S_M; all symbols are listed in the glossary)—that, repeating one of my familiar themes, have been adopted virtually exclusively within the modern scientific tradition. S_M incorporate the core methodological elements cited in the story told in the previous section, including that under them theories encapsulate the decontextualized possibilities of things, those that can be characterized in terms of the generative power of the underlying order, in dissociation from the human, social, and ecological possibilities that they might also admit. Among the decontexualized possibilities of phenomena, there are some that are identical with possibilities for technological application.

Research conducted under S_M has been extraordinarily successful: it has generated and continues to generate soundly accepted theories of a great variety of phenomena. (These theories have been the source of numerous and varied technological applications, but Kuhn has little interest in applications and his notion of fruitful or fertile research has nothing to do with success in application.) As I have said, Kuhn does not take the success of S_M to show that these theories match the ahistorical "material world." Instead, for him, it establishes that the world can be (to a marked extent) well matched to, or become amenable to grasp within, the categories of the lexicons deployed under S_M (*SFV*: chapter 7). I add that decontextualized possibilities of things are successfully identified under S_M; and, in opposition to those who adhere to materialist metaphysics, I caution that there is no reason to believe that the possibilities of things are exhausted by their decontextualized possibilities.

Within the Kuhnian picture the object of scientific inquiry is phenomena qua grasped under a strategy. Since a strategy is a methodological innovation of scientific practices, this object is not ahistorical. For Kuhn himself the aim of science is to solve puzzles whose very definition is strategy-bounded so that, in the final analysis, the questions posed in scientific inquiry are not about the "material world," but about the power of a strategy to grasp phenomena. It follows that Assumptions 1 and 2 (previous section) cannot be upheld. On my additional gloss, the aim of science is to gain understanding of phenomena, and this includes encapsulating the possibilities that they allow (section 3.2). But phenomena allow many and varied kinds of possibilities, of which not all can simultaneously be co-actualized or even co-investigated. So, actual scientific investigation opts to pursue certain classes of valued possibilities, often those valued for the sake of application, whose realizability and possibility of being investigated is historically conditioned. Thus, it is reinforced that Assumptions 1 and 2 cannot be upheld.

Fruitfulness As the Ground for Adopting a Strategy

What about Assumption 4? It could not be sustained if, for example, the (rational) grounds for adopting a strategy include that it gives rise to applications of special interest for those holding a particular value-outlook. So we must attend to the grounds for adopting a strategy.

According to Kuhn, a strategy is adopted for the sake of defining and solving puzzles, or (as I prefer to put it) for the sake of generating theories and acquiring appropriate empirical data so that theories can come to be accepted in virtue of manifesting the cognitive values highly. Then, a strategy is worthy of adoption only if it is demonstrated to be fruitful—actually to be, and continuing to be, a source of theories that come to be soundly accepted of certain domains of phenomena. A fruitful strategy, adopted in the first instance following an exemplary achievement, enables investigation to take place in the relevant field;[2] and, for Kuhn, so long as a strategy remains fruitful, research should be conducted exclusively under it. Within the scientific tradition, he maintains, fruitfulness is sufficient, as well as necessary, for the adoption of a strategy. Normally a currently fruitful strategy is in place. Then, engaging in scientific research implies adopting it—so that normally questions about adoption of strategy are neither controversial nor addressed explicitly within the scientific community[3]—until such time as the limits of its fruitful unfolding are reached. Such limits become apparent when anomalous phenomena (which have become considered important for the unfolding research) are identified, that is, phenomena that cannot, after prolonged and skillful investigation, be fitted into theories that both meet the constraints of the strategies and manifest the cognitive values highly, but at best into theories that retain empirical adequacy at the price of increasingly diminished manifestations of such other cognitive values as predictive and explanatory power and keeping ad hoc hypotheses to a minimum.

On Kuhnian views, strategies and the lexicons they bear are human creations; and a soundly accepted theory is one that succeeds in fitting certain phenomena of the world into the structured lexicon of a strategy. So it is expected that any strategy will have limits, that its fruitfulness will eventually become exhausted. (Any one kind of strategy will fail to encapsulate various kinds of possibilities of phenomena.) When the limits of an established strategy are reached, and—according to Kuhn—(allowing a certain latitude of judgment about when they are reached) only then, does the scientific tradition license the search for another strategy; and then the search is for a new strategy that can grasp the anomalies of the old one. At such ("revolutionary") moments most of the old constraints are lifted, conflicting perspectives are engaged, and there is much trial and error, until such time as a new strategy emerges (in a new exemplary achievement that offers promise of further fruitful developments) that enables the grasp of the old anomalies (cf. Hoyningen-Huene 1993: 241–43) Then that strategy comes to demand the allegiance of the scientific community.

Kuhn intends his picture to be both descriptive (under idealization) of the history of science, and normative for scientific practice. Indeed normally it does suffice for scientific research to proceed under a single strategy, provided that one accepts that what count as scientifically interesting phenomena are defined within the unfolding tradition of science, and that the aim of science is to resolve puzzles about them or to come to accept (soundly) theories of them. Proceeding in this way enables there to be successful research, practically ensures that empirical considerations will eventually lead to clear demarcation of the limits of the strategy, and keeps a measure of continuity—through the special role accorded to anomalies of old strategies—across the "revolutionary" divides that separate the periods of hegemony of succeeding strategies. Note that an argument cannot be extracted out of this that the new strategy is the only one that could have developed as successor to the old one (section 4.3). Within the Kuhnian picture, there are elements of radical contingency: that any successor at all will actually emerge; and, if one does, what its specific character will be. The emergence of a new strategy may be influenced causally by all sorts of "extra-scientific" factors (religious, metaphysical, cultural) but what matters, what legitimates the adoption of the strategy, is that it generates theories in which the anomalous phenomena can be grasped and which define new puzzles. If the aim is to solve puzzles about scientifically interesting phenomena and to introduce new ones to be solved, that is enough. Kuhn has provided a brilliant account of the transition from the hegemony of Aristotelian to that of materialist (Galilean) strategies (Kuhn 1956), as well as some less developed accounts of other "revolutionary" transitions (Kuhn 1970). Following the former transition, few products of Aristotelian science have remained in the generally accepted stock of knowledge. With the hindsight of developments under S_M (including new data, greater sensitivity to the role of certain cognitive values, and higher standards for estimating the degree of manifestation of the cognitive values in theories), it became apparent that Aristotelian physical theories were soundly accepted of very few phenomena (*SVF*: chapter 7).

Some of Kuhn's critics think that his view entails that, with the eventual anticipated surpassing of S_M as framers of research, few of its products will remain in the stock of knowledge. Thus they accuse Kuhn of a kind of relativism that seems manifestly unacceptable when we think of the discoveries of modern science and their applied successes. But Kuhn's view does not entail this. Under S_M numerous theories have been soundly accepted of countless domains of phenomena. These theories encapsulate well an increasing number and variety of the decontextualized possibilities of phenomena; and—while acknowledging the truism that empirical methodologies cannot provide certainty—there is no reason to hold that subsequent developments of the tradition will lead to removing from the stock of knowledge that these possibilities have been soundly encapsulated. Similarly, there is no reason to anticipate, for example, that the atoms of modern atomic theory will go the way of the four terrestrial elements of Aristotelian physics, at least if we consider atoms to be the constituents of molecules with capacities for generating specified effects in specified (experimental and technological) spaces. Subsequent research may lead to their refinement and elaboration but, given how soundly accepted atomic theory is, not to their rejection.

I have followed Nancy Cartwright's language here (Cartwright 1999), and her claim that established scientific knowledge is largely of capacities of objects: that they tend to have certain effects under specified (typically experimental) conditions, without the further assumption that such capacities (rather than others they might also have) will be exercised significantly in all ("natural") situations.[4] Gaining such knowledge of capacities of objects does not ground the assumption that knowledge of the "material world"—of the world as it is, independent of its relations with human beings—has been gained. Only idle skepticism would cast doubt on the existence of atoms today: there are atoms in the world "that we live in" and "that we have investigated," and we know their capacities as exercised in various experimental and technological spaces and also (no doubt) in many spaces, not of human causal origin, in which there is no (relevant) human causal involvement. Kuhn's picture fits easily with many kinds of scientific realism. But the "world that we live in" is not the "world as it is, independent of its relations with human beings."

Thus, it is consistent with the Kuhnian picture that, under S_M, we gain accumulating knowledge of decontextualized possibilities of phenomena. But, one might ask, is this really any different from accumulating knowledge of the ahistorical material world? It is, and the difference is important. In the first place, the latter idiom, unlike the former, is usually linked with the view that all possibilities of phenomena are (in the final analysis) decontextualized possibilities, and in particular with materialist reductive accounts of human cognitive (rational) and moral capacities. Secondly, the decontextualized possibilities of phenomena are those possibilities that are encapsulated by the generative power of the underlying order posited of the phenomena; they are constituted as such within scientific practices conducted under S_M. Some of them are realized in, and realizable only in, experimental and technological spaces of human creation (having been posited as the possibilities of these historically bounded spaces). Others are realized in spaces whose underlying causal order, as represented under S_M, has no relevant human involvement, but where that causal order is posited as a consequence generally of drawing upon the resources of theories accepted of experimental spaces, following scientific observation aided by instruments themselves authenticated in the course of experimental and technological practices. (In some spaces there can be good reason to hold that their possibilities are exhausted by their decontextualized possibilities.) This causal order is constituted in the course of scientific practices as a projection from experimental and technological practices; there is no basis here to infer to the features of an underlying order that is ontologically independent of human beings. (This does not prove that there is no such underlying order—Sankey 1997.)

For a theory (T) developed under S_M of a domain (D) of phenomena, "T manifests the cognitive values highly of D" implies "T soundly encapsulates decontextualized possibilities of D." These propositions become established at a particular time. Nevertheless, once established, especially if further vindicated by the success of practical applications, there is no general reason to expect that they will become vulnerable to refutation in the light of outcomes of research under different strategies, either current alternative strategies or future ones. Strategies change, and so the fundamental character of ongoing scientific investigation changes, but that permits a permanent residue of knowledge to remain, a residue that may or may not become rearticulated (as a particular case or as an approximation) under a subsequent strategy. Historicity of scientific practices does not imply the historical relativity of scientific knowledge, though it fits easily with the historical (and cultural) relativity of interest in applying particular items of scientific knowledge.

As more decontextualized possibilities become soundly encapsulated in theories, the greater is the range of technological possibilities opened up, a matter with profound social implications. For Kuhn, technological application remains principally a consequence of scientific developments, and also a source of additional empirical data to bring to bear on theories, especially by way of the instrumental partnership referred to in section 5.1. That there is widespread technological application and that it is desired are not for him among the (rational) grounds for adopting S_M; for him, those grounds are (normally) solely connected with fruitfulness, and also (at "revolutionary" moments) with being able to grasp the anomalies of the old strategy. Through this complex and subtle narrative Kuhn endorses the historicity of science: denying Assumptions 1 and 2, while retaining Assumption 4 and thus preserving an essentially internalist narrative of the history of science.

5.3 THE ROLE OF APPLICATIONS

"Application" refers to two interacting and not sharply separable roles that scientific theories can play in social life. A theory may apply *to* significant phenomena of daily life and experience; and it may be applied *in* practical activity. It applies to phenomena when it is used, by way of representing them with its categories and principles, to provide understanding of them. "Applying to" involves identifying (modeling) phenomena, as characterized using everyday categories, with phenomena as represented in the theory. A theory is applied in practical (often technological) activity when its posits inform such concerns of practice as the workings of things, means to ends, the attainability of ends, and the consequences of realizing the possible.

In Kuhn's picture, applications are important to the unfolding of the scientific tradition only as enticement for the provision of the social, material, and instrumental requirements of the conduct of research. The credibility of research depends on dissociating the value, conduct, and character of scientific practices from social and ethical evaluations of applications of the knowledge they produce.

I will now bring into play the alternative picture that I have been developing in this book, in which applications ("to" and "in") are more central than Kuhn admits. In it, particular strategies are adopted rationally (in part and in some important cases)—subject to long-term fruitfulness remaining a necessary condition of their adoption—because they can be expected to give rise to certain kinds of applications. Phenomena are in fact (and should be) brought to the attention of basic scientific investigation, not only from the scientific tradition's own unfolding (as Kuhn holds), but also from the realm of daily life and experience and social practice, from the "world in which we live." Science aims to provide understanding of phenomena and, in doing so, where appropriate, to make sense of our experience and to inform our social practices (section 3.2). Strategies worthy of adoption should normally produce theories applicable to phenomena significant for current daily life and applicable in current social practices—though normally and desirably (for substantive and methodological reasons) the reach of scientific investigation should not be limited to phenomena involved in these applications. Many significant phenomena of daily life and social practice are not fixed across historical change and cultural variety so that depending on the desired applications, different strategies may be needed. If so, Assumption 4 would not be upheld.

In order to provide detail and credibility for this alternative picture I will now develop, more fully than I have in earlier chapters, my analysis of the phenomena and possibilities of farming practices. Thence, I hope to consolidate my claim that, for some significant phenomena, competing (fruitful) strategies are possible, and that which strategies actually become adopted in research (rationally) depends upon the social locations of the investigator and upon the ways in which applications are valued from these locations. Different social locations—on the one hand, the market-oriented global economic system; on the other, grassroots movements of poor farmers—lead to the adoption of largely different (competing) strategies.

Do Materialist Strategies Suffice to Shape Research?

The modern realm of daily life and experience is unintelligible apart from the applications of knowledge gained under S_M, since it has been shaped to a great extent by identifying and realizing novel decontextualized possibilities of things. That provides a good reason for S_M to be adopted in the scientific community. The possibilities of natural phenomena encountered in daily life and social practice, however, are not reducible to their decontextualized possibilities, those they have in virtue of the posited generative power of their underlying structure (and its components), process, interaction, and law. Why, then, prioritize decontextualized possibilities in the investigation of natural phenomena? Why not attempt to create and adopt strategies under which other classes of their possibilities might be identified, for example, those they have in virtue of their places in human life and experience and social/ecological systems? Why, for example, prioritize investigating seeds so as to identify the possibilities open to them using the techniques of transgenic technology, rather than those they have in virtue of their place in productive and sustainable agroecological systems?

My answers to these questions have been given at length in earlier chapters and involve relations between S_M and the modern valuation of control (abbreviated as V_{MC}). Even so, I take it to be uncontroversial that a considerable body of scientific knowledge gained under S_M (about molecular chemistry, viral and bacterial causes of disease, soil nutrients, the components of a nutritious diet, electromagnetic radiation—to give a sample) is available to be applied in ways that can strengthen the social expression of virtually any value-outlook that is actually entertained today. This explains why it is widely valued (across value-outlooks) that scientific knowledge has been gained under S_M, and it provides a reason for the esteemed place that research under S_M has throughout the scientific community. It does not follow that research conducted exclusively under S_M (or that all research conducted under them) is valued, as distinct from inquiry in which research under S_M is balanced by (or subordinated to) research conducted under alternative strategies. That is because, in contradiction with our story, the products of research under S_M are in fact not neutral; the "evenhandedness" condition is not satisfied. Overall, and especially in fields like agricultural biotechnology in which research is dominated by specific versions of S_M, their applications favor those value-outlooks whose central practices and projects are conducted so as to further the expression of V_{MC}.

Favoring value-outlooks that contain V_{MC} violates evenhandedness because the value-outlooks of various contemporary movements and groups contest it, and theories, consolidated under S_M, do not apply to key phenomena and in significant parts of projects of importance to them. Feminist, environmental, and anti-"globalization" groups hold value-outlooks that contest V_{MC}, and also (of special interest for my argument in this book) grassroots organizations in Latin America (and elsewhere) that adopt alternative value-outlooks that emphasize such values as those of popular participation (abbreviated as V_{PP}), including local empowerment, full recognition of the entire body of human rights specified in international documents, and environmental sustainability (section 6.3). In the agricultural projects of the grassroots organizations, phenomena of sustainable productivity, preservation of biodiversity, and meeting the food and nutrition needs of the local community are all of central importance, and their practices aim to preserve and enhance productive and sustainable agroecosystems over the long haul. (Exercising control over natural phenomena is, of course, a value for them—as it is in every culture. But, unlike in V_{MC}, it is subordinated to other values, e.g., V_{PP}.) Theories developed under S_M have important applications to these phenomena and in these practices, but they are limited (or subordinated); for example, they have supplied knowledge of some of the constituents and mechanisms of agroecosystems (microorganisms, chemical nutrients), but they shed little light on the possibilities of enhancing productive and sustainable agroecosystems—in contrast, for example, to that they shed on relations between crop yields and chemical inputs to production, and on the possibilities of production with transgenic seeds. Insofar as the possibilities of enhancing productive and sustainable agroecosystems possibilities may be pertinent to desired applications, identification of them will have to be gained through research conducted under alternative strategies—agroecological strategies (abbreviated as S_{AE}).

The products of research conducted under S_M do not meet the evenhandedness condition of neutrality, for applying current knowledge gained under S_M (e.g., in agricultural biotechnology) may require conditions that would undermine the agroecological systems valued by those who contest V_{MC}; and we will see (section 9.5; section 10.4) that this is indeed the case. It remains that knowledge gained under S_M is genuine knowledge; it is expressed in soundly accepted theories or with the aid of their categories. When alternative strategies are adopted, and their results applied, that remains untouched. Consistency with soundly accepted theories is a mark of the rational; applying them in one's own projects need not be. Legitimacy of applications involves, not only that the theory has been soundly accepted, but also that its applications serve the interests of the "right" value-outlook. When a theory is applicable only in a context where certain values are expressed and embodied, to appeal to its sound acceptance as being sufficient to justify the legitimacy of applying it implies improperly limiting the range of values that may be (rationally) held. When we separate the investigations of decontextualized and other possibilities, we study things in dissociation from the conditions for the realization of their possibilities. Thus, it will not be part of the techno-scientific investigation to figure out the social conditions under which the possibilities may be realized; so we may miss that to interact with a thing so as to realize certain of its decontextualized possibilities may actually also be to treat it as a certain type of social object—for example, to realize the possibilities of transgenic seeds, we may have to turn seeds into commodities and into bearers of intellectual property rights (section 7.2).

5.4 AGROECOLOGICAL STRATEGIES

There is competition between S_{AE} and agrobiotechnological strategies (S_{BT}), which are instances of S_M. My account of agroecology is derived principally from the numerous writings of Altieri (especially Altieri 1995; I introduce some minor variations of terminology so as to fit into my general analytic framework). On biotechnology:

> In essence [biotechnology] implies the use of microbial, animal, or plant cells or enzymes to synthesize, break down or transform materials. . . . Traditional biotechnology refers to the conventional techniques that have been used for many centuries to produce beer, wine, cheese, and many other foods, while "new" biotechnology embraces all methods of genetic modification by recombinant DNA and cell fusion techniques, together with modern developments of "traditional" biotechnological processes. (Smith 1996: 2–3).

I use "biotechnology" in the sense of the "new" biotechnology. I take it to refer also to a field of scientific research, conducted under S_{BT}, that aims to produce knowledge that can enhance the methods specified in the quote. Thus, depending on context, "biotechnology" may refer either to a field of scientific research or to specific methods deployed in agricultural practices. Note that the way in which S_{AE} and S_{BT} compete does not preclude that each may draw from the positive results of the other or complement one another in limited respects (see section 6.1; section 10.3). In this chapter, since I am only concerned here with competition between agroecology (see chapter 10: note 1) and transgenic-oriented agriculture, the relevant competition between S_{AE} and S_{BT} concerns the latter only insofar as it encompasses research on transgenics.[5]

The theories that are established (soundly accepted) under the competing strategies are not inconsistent; rather they encapsulate largely different classes of possibilities which (to a significant degree) cannot be co-realized in the same cultivated lands (agroecosystems). The competition concerns which classes of possibilities to attempt to realize in agricultural practices: those of transgenics, which are of special interest where V_{MC} reigns, or those of agroecology, whose interest usually derives (in the first instance) from V_{PP}? "Technical scientific" issues pervade the competition: What is possible? What are the risks of application? Can the risks be suitably managed? But, provided that both S_{BT} and S_{AE} are fruitful, the conflict is waged largely in the realms of values, politics, economics, etc.—and where one stands in face of this conflict feeds back into the strategies one adopts in research.

Scientific Research and Ideological Critique

Agroecologists clearly recognize this; biotechnologists often do not. This is well illustrated in an exchange in *AgBioForum* between Altieri and Rosset (1999) and McGloughlin (1999) (reprinted in Sherlock and Morrey 2002). Sometimes proponents of transgenics say that the "theory" of agroecology consists simply of ideological critique, or at best of a patchwork of opportunistically gleamed fragments of traditional local knowledge. Hence the proponents of agroecology are said to be proposing not a "scientific" research program, but instead to be submitting scientific claims to ideological critique. This criticism of agroecology ignores that research under S_{AE} has been fruitful (section 10.1), so that therefore, ironically, the criticism itself is "ideological" rather than "scientific." Under both S_{BT} and S_{AE}, research aims to gain understanding of phenomena of the world and their possibilities. At the same time, it aims to gain understanding pertinent to value-laden interests in application: ". . . political determinants enter at the point when *basic* [my italics] scientific questions are asked and not only at the time when technologies are delivered to society" (Altieri 1994: 50–51). If the two kinds of aims do not appear to be "equally" in play under both strategies, this may be because inequalities of available material and social conditions enable research under S_{BT} to proceed routinely without its legitimacy constantly being called into question (section 2.1). The reasons both for and against the adoption of S_{AE}, and conversely for the virtually exclusive adoption of variants of S_M, include integrally appeal to value-outlooks. The strategies are equally "scientific," that is, held to fruitfulness that reflects commitment to impartiality, and adopted (in part) because of their relations with value-outlooks. There is not the asymmetry that critics of agroecology sometimes claim, viz., under S_M, investigation is scientific and nonideological, whereas what passes for research under S_{AE} is nonscientific and ideological (McGloughlin 1999).[6]

Those who adopt S_{BT}, misled by the myth of neutrality, tend not only to downplay the empirical achievements of agroecology and to portray it as simply ideology without link to fruitful strategies, but also to be unaware that the links of S_{BT} with V_{MC} refute the neutrality they claim for their own research. For them S_{BT} are particular instances of S_M that enable us to identify the possibilities of things (e.g., seeds) that are made available (to a significant extent) from using methods of genetic engineering. S_{BT} are indeed that; they are also those strategies whose products do and are expected to inform a particular form of technology, which is widely and almost entirely applied in practices that express highly V_{MC} (section 7.2). The first description of S_{BT} shapes research practices; the second serves to rationalize adopting them rather than other strategies.

Similarly S_{AE} have two descriptions: first, as particular instances of general ecological strategies—that frame research on the relations and interactions between organisms and their environments that enable us to identify, for example, the possibilities that things (seeds) have in virtue of their place in agroecological systems—that sometimes may be considered as *more or less* self-regulating "wholes" of which the organisms are integral parts; second, as those strategies that are intended to provide knowledge that can inform agroecological farming, which often involves embodiment of V_{PP}, for example. In both cases (S_{BT} and S_{AE}) adopting the strategies is rationalized (in part) by reference to particular values. This does not per se challenge the impartiality (sound acceptance) of the results consolidated under either strategy; it may challenge their neutrality. In the case of S_{AE}, since objects (including agroecosystems themselves) are not abstracted from their places in human experience and social relations, values enter into the subject matter of the investigation: under what conditions are certain values (e.g., those of V_{PP}) able to be further embodied? (Where S_M shape the research agenda, all comparable questions are pushed into the social science inquiries that may inform applications.) Note that the questions (posed under S_{AE}) are about the degree of embodiment and manifestation of the values; reaching empirically based results about them (as distinct, perhaps, from having an interest in them) is logically independent of endorsing them. There can be impartial results about the degree of manifestation and embodiment of values (section 3.1; chapter 11).

Research Conducted under Agroecological Strategies

It is in virtue of the first description that S_{AE} shape research practices. I will elaborate a little. Under S_{AE}, research aims to confirm generalizations concerning the tendencies, capacities, and functioning of agroecosystems, their constituents, and relations and interactions among them. These include generalizations in which, for example:

> "Mineral cycles, energy transformations, biological processes, and socioeconomic relationships" are considered in relationship to the whole system; generalizations concerned not with "maximizing production of a particular system, but rather with optimizing the agroecosystem as a whole" and so with "complex interactions among and between people, crops, soil and livestock." (Altieri 1987: xiv–xv)

To illustrate:

> Low pest potentials [are likely] in agroecosystems that exhibit the following characteristics: high crop density through mixing crops in time and space; discontinuity of monocultures in time through rotations, use of short maturing varieties, use of crop-free or preferred host-free periods . . . ; small, scattered fields creating a structural mosaic of adjoining crops and uncultivated land which potentially provides shelter and alternative food for natural enemies . . . ; farms with a dominant perennial crop component . . . ; high crop densities or the presence of tolerable levels of specific weed species; high genetic density resulting from the use of variety mixtures or crop multilines. (Altieri 1999: 24–25)
>
> Restoration of natural controls in agroecosystems through vegetation management not only regulates pests, but also helps to conserve energy, improves soil fertility, minimizes risks, and reduces dependence on external resources. (Altieri 1994: 150)

Of particular salience are generalizations that help to identify the possibilities for productivity and sustainability of agroecosystems, where "sustainability" has been defined in terms of four interconnected characteristics:

> Productive capacity: "Maintenance of the productive capacity of the ecosystem"; Ecological integrity: "Preservation of the natural resource base and functional biodiversity"; Social health: "Social organization and reduction of poverty"; Cultural identity: "Empowerment of local communities, maintenance of tradition, and popular participation in the development process." (Altieri et al. 1996: 367–68)

Theories, under S_{AE}, may be considered to be constrained so as to be able to represent sets of generalizations of the above kinds (*SVF*: 193–96) and the hypotheses (drawn from general ecological theory) that are entertained for their explanation and determining the limits of their application. The generalizations of agroecology tend to express probabilistic relations or tendencies, and they may have greater or less specificity. So, discerning the limits of application of these generalizations is especially important. Note how the generalization: "Enhancement of biodiversity in traditional agroecological systems [in Latin America] represents a strategy that ensures diverse diets and income sources, stable production, minimum risk, intensive production with limited resources, and maximum returns under low levels of technology," is later qualified by: "We have still not been able to develop a predictive theory that enables us to determine what specific elements of biodiversity should be retained, added, or eliminated to enhance natural pest control [and other desired outcomes]" (Altieri 1994: 7, 38).

Participants in Agroecological Research

Data are selected and sought out in virtue of their relevance for appraising these theories and for enabling phenomena, relevant in the light of, for example, the values of V_{PP}, to be brought within the compass of a theory's applicability. Obtaining the data often requires subtle, regular, painstaking, accurate observation and monitoring of a multiplicity and heterogeneity of details in the agroecosystems. The skills for this are usually only developed by local farmers themselves, so that obtaining the data depends on the collaboration of local farmers and the utilization of their experience and knowledge, and the lexicon in which they are reported will reflect the distinctions and categories of this experience. Agroecology cannot be pursued with a sharp distinction between the researcher and the farmer; the farmer's observations are essential to the conduct of the research:

> Seeds have multiple characteristics that cannot be captured by a single yield measure, as important as this measure may be, and farmers have multiple site-specific requirements for their seeds, not just controlled condition high-yields. . . . The

inescapable conclusion is that a different approach, participatory breeding by organized farmers themselves, which takes into account the multiple characteristics of both seed varieties and farmers, is essential. (Rosset 2001; cf. Machado and Fernandes 2001)

Quantitative data are often pertinent: counting the number of pests in a given area, measuring the size of crop yields, amount of water available, etc.; statistical comparisons—for example of pest populations across monocultures and polycultures, or of the yields of different crops when different methods are used. Experimental data are sought both to support statistical comparisons and to demonstrate that possibilities can be realized in agroecosystems with certain characteristics, for example: "It is possible to stabilize the insect communities of agroecosystems by designing and constructing vegetational architectures that support populations of natural enemies or have direct deterrent effects on pest herbivores" (Altieri 1994: 7). In agroecological contexts, an "experiment" involves introducing, for the sake of observing its systemic effects, a modification (under an investigator's control) of an agroecological system. Given the local distinctiveness of agroecosystems, the mark of a "good" experiment cannot be its replicability across diverse environmental and social conditions. Note that control is involved in agroecological experiments and farming practices, but subordinated to the values, for example, of V_{pp}.

Continuity with Traditional Local Knowledge

Relevant data are often obtained from the study of farming systems in which traditional methods informed by traditional local knowledge are used. These systems are appropriately submitted to empirical scrutiny because agroecological studies have shown "that traditional farming systems are often based on deep ecological rationales and in many cases exhibit a number of desirable features of socio-economic stability, biological resilience, and productivity" (Altieri 1987: xiii; see Altieri 1995: chapter 6 for details and examples). They exemplify many known agroecological principles, and, one can expect, others will be extracted in the course of studying them (Altieri 1995: 143). They can (with adaptations suggested by research findings) be enhanced with respect to all four of the characteristics of sustainability listed above; and, especially with respect to "cultural identity," they are often uniquely appropriate for the activities of poor, small-scale farmers. It is worth noting that the methods used in these systems have been tested rigorously in practice, and have been particularly effective (reflecting the experimental approach of traditional farmers) over the centuries in "selecting seed varieties for specific environments" (Altieri 1995: 116); these are often the seed varieties (or the original sources of them) that are modified genetically to produce transgenic varieties (Kloppenburg 1987).

5.5 ADOPTING A STRATEGY AND THE SOCIAL LOCATION OF SCIENTIFIC RESEARCH

At least in some fields, there can be multiple strategies that compete in the way described above; and each of the competing strategies may be fruitful. Then there arises the question of which strategy to adopt, one for which different answers may be proposed and acted on by different investigators. I have suggested that actual answers (explicitly or implicitly) often draw upon mutually reinforcing relations between adopting strategies and the value-outlooks whose interests would be served especially well by applications of knowledge gained under the strategies.

There is, however, a general reason (soundly based on cognitive interests that are grounded in the aim of science stated in section 3.2) in favor of developing research under some strategies other than S_M: to test whether all possibilities can become grasped under S_M. By identifying possibilities that are not identical with possibilities currently encapsulated by soundly accepted theories under S_M, we can pose concrete challenges for research conducted under S_M to meet. This reason would not appeal to Kuhn; he holds that such challenges are unnecessary since, in due course, anomalies will accumulate in the normal unfolding of S_M. However, there may be bounds to S_M, even though within the bounds there remains an unlimited set of possibilities to be identified. Only tests of the kind indicated here can hope to identify these bounds. (I am not sure that Kuhn recognized this.) By identifying possibilities of the kinds indicated it could be probed empirically whether or not there are bounds to the development of S_M.

Pluralism of Strategies

This general reason for multiplying strategies sits in tension with the rationales for adopting specific strategies, especially when we remember that there is also competition for the resources needed to conduct research. Resources devoted to probing the limits of S_M in this way may be resources taken away from pursuing more favored projects and, de facto, giving the resources to support research whose strategies gain their primary rationale from competing value-outlooks. Only a satisfactory resolution of this tension, I believe, could restore neutrality as a compelling value of scientific practice. The tension is heightened when we consider the legitimacy of applications.

Kuhn's picture, recall, portrays applications mainly as consequences of scientific developments; and developments under S_M have identified numerous decontextualized possibilities that have become, and are continuing to become, realized in applications at an increasingly rapid rate. The efficacy of applications depends on the input of sound scientific knowledge that can be provided (for many applications) by research conducted under S_M. The legitimacy of some applications depends also on claims about the possibilities of things (section 3.4; section 6.4; chapter 10). Consider the claim: under S_M, means (involving developments of transgenics) may be identified for producing food sufficient in quantity to continue to feed the world's population. Assuming that this is true (in fact it is contested), applying the knowledge thereby obtained would be legitimated, if there are no "better" ways of producing sufficient amounts of food—ways, for example, that would be part of agroecosystems that were structured so that the food is not only produced but also so that everyone is actually fed sufficiently and nutritiously, and that sustainable (and improving) and productive agroecosystems are maintained (chapter 10; Altieri and Rosset 1999; Kloppenburg and Burrows 1996). But the possibility of producing sufficient food by developed and expanded uses of agroecological methods cannot be investigated under S_M (section 10.3). So research conducted under S_M cannot provide a crucial item of knowledge (or the means for attempting to gain it) needed to justify affirming that the use of transgenics is essential to the solution of the world's food problems. (A similar point can be made in connection with investigating risks—chapter 9.)

The proponents of the intensive use of transgenics in agricultural production respond that there is no evidence that developed agroecological methods could produce sufficient food. In responding to this assertion it is important to keep in mind that producing sufficient quantities of food to feed everyone does not imply that everyone will be fed. Currently, sufficient food is produced worldwide, but hunger persists (Boucher 1999). Given that agrobiotechnology plays an integral role in the global economic system, under which hunger currently persists, one might wonder why the expansion of production of food promised by the new methods will be any more likely to lead to the hungry being fed. Who is fed, and who is not, is not independent of the methods of production. Even if there were strong evidence that enhanced agroecological methods could not produce sufficient food to feed everyone, it might still be the case that agroecology needs to be developed so that the currently hungry and their descendants will be fed. The legitimacy of the furthering of agroecology needs only this more modest claim. Perhaps, in order that everyone is fed, a variety of farming methods will have to be used. The issue is an empirical one, but investigations conducted exclusively under S_M cannot adequately address it. It can only be responsibly investigated within a theoretical framework that investigates the full causal nexus of production and consumption of food, and more generally of human well-being; and in a process that is responsive to the needs, interests, and value-outlooks of everyone (chapter 10).

As things stand, decisive evidence is lacking that agroecological methods can be enhanced and expanded to produce sufficient food to feed everyone (section 10.4). However, that could be because, while S_{AE} have displayed a measure of fruitfulness (section 10.1), their limits have effectively been reached; or because, due to lack of the necessary social conditions and material resources, there has been much less research conducted under S_{AE} than under S_{BT}. This matter could be explored empirically by providing conditions to further develop agroecology in those areas where there is hunger and an available rural workforce (thus furthering it under the legitimacy of the modest claim referred to in the previous paragraph, in areas where its effectiveness has been repeatedly demonstrated— Altieri et al. 1996); this would enable virtually risk-free investigation of the possibilities of agroecological production. (Cf. section 8.3).

The proponents of transgenics see little urgency in conducting such investigation because, I think, the widespread implementation of agroecological methods would be incompatible with the social structures, values, and policies under which transgenic technology is developing. Perhaps, in the final analysis, they hold that agroecological methods cannot produce sufficient food, because these methods cannot be developed under current social conditions and so because they are irrelevant in the current socioeconomic climate (section 10.6). Hence, perhaps the proponents hold that there is no better way to produce the needed food, because transgenic technological methods are confirmed as providing the most efficacious of the available possibilities whose realization could be informed by theories established under S_M, and thus could most usefully further the expression of V_{MC}; and, for them, furthering V_{MC} (and so, at the present moment, fitting into the market-oriented global economic project) has become a condition on a legitimated way. If this is so, then the legitimacy they offer for prioritizing the use of transgenics in agriculture does not rest upon empirical confirmation that agroecological methods are insufficient for producing the food (and so it does not beg questions, which to be addressed empirically would require developments of S_{AE}). It rests upon commitment to V_{MC} (and the values of market-oriented "globalization"), upon valuing the decontextualized possibilities identified by research on transgenics because to realize them is, at one and the same time, to realize valued social possibilities (section 2.1; section 10.6; section 11.1).

There are no "scientific" reasons—reasons based on the cognitive aims of science—to decline to appraise empirically the possibilities of agroecology. Given that applications involve issues not only of efficacy, but also of legitimacy or social value, it is just arbitrary to insist that what counts as a "scientifically" interesting phenomenon is determined only in view of the internal unfolding of the scientific tradition, and not also by interests connected with application. Thus, the competition between S_{BT} and S_{AE} cannot be dissolved by appealing to the general character of science. The marginalization of S_{AE} in the mainstream, I have suggested, is explained (when we probe for the reasons) not because, after adequately providing for efforts to develop them, serious doubts about their fruitfulness have been confirmed. Rather it is because they cannot lead to applications of interest for V_{MC}; and perhaps also because, if their fruitfulness were confirmed, the legitimacy of prioritizing the use of transgenics in agriculture would be challenged—though in fact the proponents of transgenics tend not even to entertain that the far-reaching fruitfulness of S_{AE} might, given the opportunity, be confirmed.

Conversely, the reasons for adopting S_{AE} (which, I repeat, draw in many ways upon basic knowledge gained under S_M in all sorts of ways) as an alternative to S_{BT}, are connected with challenges to V_{MC} and with holding such competing value-outlooks as that of V_{PP}. Either way adoption of strategies, and thus the character of research conducted, is unintelligible if dissociated from the social location of scientific practices and their applications; and thus, in turn, social location can serve as a ground (but not one that downplays the importance of fruitfulness) for critique of scientific practices, and as a source and condition of alternatives.

5.6 SOCIAL LOCATION AND ADOPTING A STRATEGY: ADDITIONAL SUPPORT FOR PLURALISM

The objects of scientific inquiry are phenomena as grasped under a strategy, so much so that they vary with strategies and cannot be characterized in strategy-neutral terms. That is Kuhn's insight. Strategies, a key component of scientific methodology, are historically variable, and so too are the objects of scientific inquiry.

To understand phenomena is to describe and explain them and to identify the possibilities they admit. A strategy has the resources to identify a particular class of possibilities. Any one strategy is worthy of adoption only if, given the opportunity and appropriate resources, it shows itself to be fruitful, that is if it is successful in actually identifying possibilities of the relevant class (encapsulating them in soundly accepted theories). Competing strategies explore classes of possibilities (often of the same phenomena, e.g., seeds) that cannot be co-realized; for example, realizing (to any significant extent) the possibilities of transgenic seeds cannot be co-realized with key possibilities of productive and sustainable agroecosystems (chapter 7).

When two fruitful strategies compete, what are the reasons to adopt one of them rather than the other? Since both are fruitful, reasons based exclusively on cognitive (epistemic) value cannot favor one rather than the other. Elaborating the model of theory choice proposed in section 4.3, my answer is: Adopt the one that enables us to gain understanding that is applicable to phenomena and (where appropriate) in practical projects of significance for our value-outlooks, thus the one that identifies possibilities that, if realized on application, would further these projects. This provides a good reason to adopt a strategy without, at the same time, denying that the scope and value of the basic understanding gained in scientific research transcend interest in applications. It is a reason that points to the (social) value of research conducted under the strategy.

Both fruitfulness and applicability are among the grounds for adopting a strategy. We adopt a strategy partly for the sake of consolidating theories that are applicable in ways that are significant for our value-outlooks. In a particular field of research, there may be no relevant disagreements across value-outlooks about what are the phenomena and projects for which applications of theories are desired. Then competing strategies are unlikely to emerge. Different value-outlooks, however, may (in some fields) lead to different appraisals of the significance (social value) of applications, and thus to their respective adherents adopting competing strategies (e.g., S_{AE} and S_{BT}). Where this happens a case can be made that a plurality of strategies should appropriately be supported within the whole scientific community (despite the resulting tensions that would be occasioned by the fact that the classes of possibilities being explored are not co-realizable in the same contexts—section 4.1). Moreover, if my analysis is correct, it will be no surprise—when there is a hegemony of values (e.g., V_{MC} or those of the global economy) in the scientific community and its supporting institutions—that one kind of strategy comes to be adopted virtually exclusively in the scientific community and that adopting that strategy is not generally recognized as a matter of choice or as in need of rational support.

According to the picture I have offered, application plays a central role in shaping scientific practice. It is not just a consequence (or instrumental partner) of successful research but it, where it is valued in social practices that one endorses, is part of the very reason to adopt a strategy. We might put it: Possibilities, insofar as they are identical to possibilities for application, partly constitute the object of scientific inquiry. Such possibilities, of course, are objects of social value and historically and culturally variable, functions of the social location of the scientific practices. The strategies we adopt are those suitable for exploring these possibilities, and so they too must vary as a function of the social location of scientific practices. Thus, applications—successful, desired, anticipated—feed back so as to influence at the most fundamental methodological level the way in which scientific investigation is conducted.

NOTES

1. Under certain conditions connected with applications (next section), theory choice may be made across strategies (*SVF*: chapters 7 and 10; cf. section 4.3).

2. I will not keep repeating the qualification: "in the relevant field." It applies to all the remarks about strategies (and research framed by a strategy) that follow throughout the book.

3. This is a factor (additional to those discussed in section 2.1) that helps to explain why often it is thought that science is just that inquiry conducted under the currently dominant strategies. (Modern science is inquiry conducted under S_M; and professional formation apprentices

the newcomer into such inquiry—in institutions where, for the most part, V_{MC} is not challenged.) Since normally there is no controversy about S_M in the scientific community, their role can easily remain hidden so that it is not recognized that there may be other strategies, and thus investigation conducted under another strategy tends to be dismissed as "unscientific" (section 5.3; section 4.3; chapter 7).

4. This point is the key to criticisms of behaviorist (and other) approaches to experimental psychology that I have developed in several articles, for example, Lacey and Schwartz (1986; 1987). On the issues under discussion here I have been influenced by writings of Roy Bhaskar, for example, Bhaskar (1975).

5. In section 4.1 I discussed the relationship of the kind of competition, exemplified by that between S_{AE} and S_{BT}, with what Kuhn has called "incommensurability." See the paragraph after the next for how it involves the difficulties of communication that Kuhn diagnoses to be part of incommensurability.

6. She writes: "Altieri and Rosset's arguments are neither scientifically supported or even really about biotechnology. Their arguments are primarily directed against Western-type capitalism and associated institutions (e.g., intellectual property rights, the WTO). Biotechnology is used as a Trojan Horse." (McGloughlin 1999: concluding comments).

Part II

CURRENT CONTROVERSY ABOUT TRANSGENIC CROPS

Chapter Six

The Controversy about Transgenics: Structure and Opposing Interests

The development and practical utilization of transgenic plants—like developments in medical biotechnology and in communications and information technologies—is among the notable recent advances of techno-science. It is shaping rapid and far-reaching changes in farming practices in several parts of the world and, at the same time, meeting strong resistance from a variety of groups and interests.

Transgenic plants are grown from seeds that have been "genetically engineered." Genetic materials, often taken from organisms of unrelated species, have been inserted into their genomes, using the techniques of recombinant DNA, in order that the mature (or still growing) plants acquire specified "desired" properties, such as herbicide or pesticide resistance and toxicity to certain classes of insects, or in order that their products become sources of nutrition. The techniques of genetic engineering enable genetic modifications of plants that would not occur by means of the mechanisms of natural selection or the crossbreeding methods used by farmers and conventional plant breeders. Mae-Wan Ho sums it up nicely:

> Genetic engineering is a set of laboratory techniques for isolating genetic material from organisms, cutting and rejoining it to make new combinations, multiplying copies of the recombined genetic material (recombinant DNA) and transferring it into organisms, bypassing the process of reproduction. Genes can be exchanged between species that would never interbreed in nature. Thus, [daffodil genes end up in rice ("golden rice")] and bacterial genes in plants. (Ho 2002c)[1]

At the present time, the transgenics (TGs)[2] most commonly used in agriculture are of two types: those engineered to contain genes that create resistance to herbicides containing *glyphosate* (such as Monsanto's RoundUp); and those engineered to contain a gene from the bacterium Bt (*Bacillus thuringienis*) that makes the growing plants release a toxin that functions as a pesticide. Varieties of each have been developed for several crops, including corn, soybean, canola, and cotton.

6.1 CONTROVERSY ABOUT THE LEGITIMACY OF USING TRANSGENICS

Controversy has mushroomed during the past fifteen years about the development and increasingly widespread planting of TG crops and marketing of their products. It is not about the *efficacy* of currently used TGs; it is not contested, for example, that RoundUp Ready plants really are resistant to glyphosate, and Bt plants really are toxic to certain kinds of insects, and that the range of efficacious TG technologies can be vastly expanded (but see section 6.4 below). But, apart from this there is little agreement and little common ground to draw upon. In large part this is because the following questions remain poorly articulated: Exactly what is at stake? What knowledge (and lack of knowledge) is relevant to the debate? Who are the bearers of relevant knowledge? Is the debate primarily about scientific matters or socioeconomic ones, or both together with (perhaps) ethical and religious ones? Does TG technology raise special concern or is the debate really (if covertly) about all techno-scientific innovations? The lack of clarity surrounding these questions reflects that fundamentally opposed values, interests, and ways of life are at stake. At the same time, however, questions about scientific results and interpretations of their significance, about what needs to be investigated scientifically, and even about the character of scientific investigation itself, are in dispute. The resulting interplay of values and science can lead to a missing of the minds and to an impasse, sometimes marked by name-calling, law suits, and disruptive confrontations, from which there seems to be no constructive and reasoned way out. Despite this, I will show that there is an intelligible structure underlying the controversy. It is greatly illuminated by drawing upon the conclusions of part I of this book, and it enables us to consider the merits of the various positions and the possibilities for reconciliation and compromise, to discern how scientific research might be brought to bear effectively on disputed issues, and to discuss how agriculture (and what varieties of it) should be practiced in a democratic society.

A sense of urgency marks the controversy. Plantings of TG crops have expanded rapidly and exponentially and, right now, large agribusiness corporations (supported by the policies of a growing number of governments) are engaged in the project of ushering in the intensive and more widespread use of TG crops, aiming thereby to shape the agriculture of the future. Indeed, the controversy exists because agribusiness has introduced and gone ahead with this project in the pursuit of its own interests, neither (according to the critics) in response to a scientific consensus that using TGs is urgent, risk-free, and indispensable, nor after having gained a popular democratic mandate. The critics are trying to stem this tide; the proponents want to consolidate and expand their gains. If they cannot make their case now, the critics sense, either there will be a worldwide agricultural catastrophe or it will be too late to change the course set by the forces favoring the use of TGs. On the other hand, the proponents are seizing an opportunity to extend (what they consider) economic and scientific rationality into a domain of social life (agriculture) that has hitherto been resistant.

Interpreting the Controversy

The controversy is about *legitimating* (or not) research, development, practical agricultural implementation of TGs, and practices and policies (pertaining to TGs) that currently are being put into effect under the sponsorship principally of agribusiness corporations. It is fruitful to represent the main dispute as one in which the proponents (the P-side) argue for the legitimacy (and importance) of the development, immediate implementation, intensive utilization, and widespread diffusion of TGs, in the agricultural practices that produce major crops, throughout the world as soon as possible, and for this to become a central plank in public agricultural policies. Opposing this, the critics (the C-side) deny that the P-conclusions have been adequately established; they argue that more research is needed before a definitive position can be taken; and, positively, they prioritize alternatives (such as agroecology) that do not use TGs, and the urgency and priority of investigating their productive potentials. The C-side is compatible with the eventual acceptance of the use of TGs (at least of some of them in some circumstances); it is not opposed to the use of some biotechnological techniques (that do not involve use of TGs) in agriculture, for example, genomic analysis to help selection of new varieties of crops; and it may support the continuation but not the priority of research and development of TGs, and even their immediate small-scale use for addressing specific problems for which other solutions are not currently known.

In order to resolve the dispute between the P- and C-sides, or to find out if there are insuperable obstacles to bringing about a resolution, it is necessary to identify clearly what lies behind their opposed conclusions. To this end, I encapsulate the controversy: first, by identifying four pairs of contrary propositions ($P_1/C_1-P_4/C_4$) that are in dispute (section 6.2)—each of the pairs will be discussed, in turn, in a chapter by itself; and, second, by sketching the value-outlooks that are implicated, respectively in the two positions (section 6.3). Encapsulating the dispute in this way I hope to display a perspicuous contrast[3] between the two sides, one that meets the following conditions: (i) each side can acknowledge that its position has been fairly represented; (ii) each side becomes able to recognize the internal coherence of the other, to identify clearly what lies behind the disagreements and to raise questions about the evidence and arguments that support the various propositions; (iii) avenues that might lead to resolution, that are consistent with the basic commitments of each side, become opened for exploration.

In the public domain, of course, reasonable resolution might not be possible, perhaps because one side has a monopoly of political and economic power and it sees no need to compromise with the arguments and values of others. In any case, if the goal is only to "win" the dispute, there will be little interest in understanding the opposed side, as distinct from portraying it so as to justify dismissing its claims; then, my attempt at interpretation may be dismissed by both sides as de facto providing support for the other side. But the proponents of TGs regularly present themselves as having the support of the authority of science. I offer my interpretation of the controversy in the hope of being able to test this claim. This requires gaining understanding of both sides of the controversy; and to do this one needs a language—for presenting a perspicuous contrast—that expresses how scientific practices and social/ethical/political/economic values may interact, how values play a role in scientific practices, and how value judgments may be influenced by the outcomes of scientific inquiry. The viewpoint on these matters, developed in part I, provides us with this language.

Limits of the Discussion of Part II

It is fruitful to portray the controversy as between two well-defined contrasting positions. But I do not wish to give the misleading impression that, in the public controversy, all the proponents and critics share respectively positions that I lay out as the P-side and the C-side. In the appendix, I indicate the sources that I have used in constructing my portrayal (appendix: 1), but there are plenty of viewpoints that do not line up neatly with it. There are more moderate and nuanced positions: for example, those that question some but not other uses of TGs; those that couple claims about a farmer's right to plant and market TGs with upholding consumers' right to know the ingredients of the food they buy; and those that just express caution and want developments to slow down. There are also more radical positions, including those that reject any use of TGs out of hand (sometimes on religious or radical ecological grounds). I will review some of these variant stances (appendix: 1, variants), but they will not play a major role in my discussion. Nevertheless, no matter what variant one upholds, it is difficult to avoid some commitment regarding the substance of each of the four pairs of propositions, perhaps not one of the extreme positions I state, but a more moderate one or some sort of compromise. We can, I think, represent most viewpoints in the controversy in terms of how they accord with or depart from one of the items of each pair.

One final note of clarification, I address only issues pertaining to using TG plants for major agricultural crop production. Thus, I do not represent the C-side as taking a position against agricultural biotechnology in general (only against multiplying large-scale uses of TGs at the present time); the types of agriculture that it favors may, for example, use genomics as an aid to chart ecosystems, to identify the vulnerabilities of pests, or to develop techniques of bio-sensing for identifying toxic waste or water pollution; and it may use certain biotechnological techniques (e.g., tissue culture) to improve the reproduction of plants that may be important for sustainable agroecosystems (Guerra et al. 1998a; 1998b), or knowledge of the genomes of crop plants as an aid to farmer-selected breeding (Miguel Guerra, personal communication, March 2, 2004). I also do not address other kinds of uses of TGs, for example, their uses for producing pharmaceutical products (insulin in pigs, vaccines in bananas) or for studying basic biological processes (protein formation, developmental studies). There is controversy about some of these uses too, but it is outside of the scope of this book.

6.2 SUPPOSITIONS OF THE PROPONENTS' AND CRITICS' ARGUMENTS

The P-side, drawing upon the prestige of techno-science, maintains that TG technology is efficacious, beneficial, legitimate, and even has an obligatory place in national agricultural and trade policies. The C-side, prioritizing such values as environmental sustainability and the empowerment and well-being of communities of small-scale farmers, challenges the value of the alleged benefits and the legitimation of using the technology, and discerns greater promise in alternative approaches to farming such as agroecology. The following four pairs of contrary propositions capture well the disagreement.

The Contrary Suppositions

Not surprisingly, in light of the analysis of part I, the different stances taken towards techno-science, as well as the values held by the respective protagonists, have impact on what are considered the appropriate strategies to adopt in scientific research pertaining to agricultural issues and, therefore, on what kinds of knowledge one expects to apply in order to enhance the efficacy and valued outcomes of farming practices.

Strategies for Research in Agricultural Science

P_1 Developments of transgenics are informed in an exemplary way by scientific knowledge, that is, they are informed by knowledge gained in research conducted under appropriate versions (biotechnological) of materialist strategies; they are instances of techno-scientific developments, which are the principal sources of improvements of agricultural practices and (more generally) meeting human needs.[4]

C_1 The kind of knowledge gained under materialist strategies is incomplete and cannot encompass the possibilities of, for example, sustainable agroecosystems and the possible effects of uses of transgenics on the environment, people, and social arrangements; it is necessary to adopt other strategies in order to investigate these matters.

Clearly TG crops would not be planted, and their products marketed, processed, and consumed, if they were not perceived by their users to be sources of benefits, and, of course, one who considers them to be beneficial will be reluctant not to support their use.

Benefits of Using TGs

P_2 There are great benefits to be had from using TGs now, and these benefits will greatly expand with future developments, among which are promised TG crops with enhanced nutritional qualities that can readily be grown in poor developing countries so that TGs may become key to addressing problems like those of hunger and malnutrition. When these promises are fulfilled, the benefits of TGs will become spread evenhandedly so as (in principle) to serve the interests and to improve the farming practices of groups holding any viable value-outlooks.

C_2 The benefits claimed for currently used TGs reflect the ethical/social values of agribusiness, large-scale farmers, and others who are beneficiaries of the global market. Furthermore, not only are the benefits relatively slight (perhaps even exaggerated by the proponents), being confined largely to these groups and not extending to small-scale farmers in the "developing" world (or to organic farmers in the advanced industrial societies), but also the promises made about future benefits are not credible, in part because developments of TGs reflect the interests of the global-market system, the very same system within which poverty, the fundamental cause of hunger and malnutrition, persists today.

Benefits alone do not suffice for legitimation. It requires also that the uses of TGs must not cause or occasion the significant risk of causing serious harm.

Risks of the Development and Use of TGs

P_3 There are no hazards to human health or the environment arising from the current and anticipated uses of transgenic crops and their products that pose risks—of seriousness, magnitude, and probability of occurrence sufficient to cancel the alleged value of their benefits—that cannot be adequately managed under responsibly designed regulations.

C_3 This claim about risks is not well established scientifically. Moreover, the greatest risks may not be direct ones to human health and the environment mediated by biological mechanisms, but those occasioned by the socioeconomic context of the research and development of transgenics and their associated mechanisms, such as designating that transgenic seeds are objects to which intellectual property rights may be granted.

Those who uphold P_2 and P_3 (and the adequacy of current regulatory mechanisms) may hold that this is sufficient to ground the right of farmers to use TGs should they so desire (and some proponents seek no more legitimation than this): "I judge them to have benefits for me and they occasion no serious risks (so that there is no reason to think that using them will harm anyone else), so I am entitled to use them." Remember, however, that the P-side argues for (and the C-side against) the intensive and more widespread use of TG crops in present day and future agriculture and for policies that support such use. Even if P_2 and P_3 do ground the right of farmers to use TGs, they do not suffice to ground public agricultural policies in which uses of TGs are prioritized. (Public policy might also lead to qualification of the alleged right, or to introducing conditions that would devalue its exercise.) Legitimating that using TGs be prioritized would depend on there being no other forms of farming with comparable or greater benefits and with comparable or lesser risks. And, if there are such alternatives, those who endorse C_3 point out, the mechanism for controlling risks would have to take into account whether or not using TGs on a wide scale would threaten the conditions on which the alternatives depend. The fourth pair of propositions deals with this matter.

Alternative (or "Better") Forms of Farming

P_4 There are no alternative kinds of farming that could be deployed instead of the proposed transgenic-oriented ways without occasioning unacceptable risks (e.g., not producing enough food to feed and nourish the world's growing population), and that reasonably could be expected to produce greater benefits concerning productivity, sustainability, and meeting human needs— "transgenics are necessary to feed the world."

C_4 Agroecological methods (and other alternatives) can be and are being developed that enable high productivity of essential crops (and occasion relatively less risk); and they promote sustainable agroecosystems, utilize and protect biodiversity, and contribute to the social emancipation of poor communities. Furthermore, there is good evidence that they are particularly well suited to ensure that rural populations in "developing" countries are well fed and nourished, so that without their further development current patterns of hunger are likely to continue.

P_4 does not imply that TG-oriented agriculture by itself is sufficient, but only that, if the world is to be fed adequately in the not-too-distant future, TGs will have to play a major role, for there is no alternative kind of farming that can be used, where TG-oriented ways are proposed to be used, to achieve this end. C_4 challenges this; or it may be interpreted as denying that P_4 is evidentially well supported and affirming that the preponderance of evidence points towards C_4. In addition it suggests that developments of alternative methods, for example, agroecology, are necessary to make it possible for the currently impoverished (and their descendants) to be fed. As I have stated C_4, it does not rule out that there may be some role for TGs in the agriculture of the future. (Some radical critics deny this—see appendix). Note that important variants of P_4 will be considered in section 10.6.

Interpretation and Taking a Neutral Position

In the previous section I listed three conditions that I expected my interpretation to meet. I am confident that, by grounding it in the four pairs of suppositions, it does meet those conditions. But it does not follow that the interpretation is neutral between the two sides. Anticipating, I think that there is good reason to endorse C_1 (methodological pluralism),[5] that *now* P_3 (about risks) and P_4 (about alternatives) lack the support that they need to have to play their role in arguments legitimating uses of TGs, and that there is urgency to conduct research relevant to testing the limits of the promise of alternative agricultural methods, expressed in C_4. That is enough to deny legitimacy *at the present time* to projects aimed at the widespread implementation of TG-oriented agriculture throughout the world. In this sense, the proponents of TGs are right that my interpretation (at least when combined with my assessments of the current extent of evidential support for P_3 and P_4—which a proponent might attempt to rebut) provides support for the opposing side—for they do not want to concede that there are good reasons to delay further innovations of TG technology. But it does not necessarily deny legitimacy to these projects *in the long run* (and some critics have considered this to be providing support for the proponents). As we will see, the legitimacy of the TG project in the long run depends on the outcomes of testing the limits of C_4. My interpretive framework does not guarantee that the opponents of TGs will be vindicated in the long run; it sets up a context in which empirical investigation, conducted under a plurality of strategies (including agroecological ones), could play a major role in cutting through the disagreements about risks (P_3/C_3) and alternative types of farming (P_4/C_4).

6.3 THE VALUE-OUTLOOKS OF THE PROPONENTS AND THE CRITICS.

At the beginning of section 6.1, I said that fundamentally opposed values, interests, and ways of life are at stake in the controversies about TGs. Not only matters open to scientific investigation (as well as matters about the character and forms of scientific investigation) but also matters concerning values are at play in the appraisal of the four pairs of propositions. This is partly why the controversies seem to be so intractable and why so often each side misinterprets, and harshly judges, the other. The P-side, comfortable carrying the mantle of "science," often dismisses the C-side as merely representing ideology, standing in opposition to science and to progress. On the other hand, the C-side claims to be on the side of "ethics." It tends to see the P-side, not so much as an agent of science, but as fostering the "biotechnological revolution" in agriculture for the sake of the political-economic project of "globalization" (see section 10.6); and (informed by C_4) it maintains that, within the structures of globalization, the food, nutritional, and other needs of the poor cannot be satisfied. This, when combined with C_3, leads to the charge that P_2 and P_3 are supported only under the cover of lies, deception, and corporate bullying. The language deployed can be harsh: the introduction of TGs is ethically outrageous, and the behavior of the corporations that produce them (and the government policy-setting that supports them) should be condemned as grossly unethical, since (it is said) it risks disease, environmental damage, the livelihood of small agricultural producers and their communities, even the integrity of the world's food supply and democratic institutions, all for the sake of the expansion of their dominance of the whole agricultural process and their own profits. Are such harsh ethical judgments warranted? Or should we interpret the behavior of those on the P-side as expressing a different ethical value-outlook to that held on the C-side? Partisans of the P-side often claim that their outlook is more appropriate for responding ethically to the realities of the contemporary world than the outlook of the C-side, which they charge represents the frustrated or nostalgic gasps of a value-outlook that the advances of techno-science have rendered obsolete? Before trying to cut through such charges and countercharges, I will digress in order to insert some summary remarks about ethics. (These remarks should be read in conjunction with the remarks on values in section 1.1 and section 3.1. The arguments for them can be readily extrapolated from the arguments in these passages. See also *SVF*: chapter 2; Lacey and Schwartz 1996).

Ethical Values

Ethical values concern, first and foremost, the constituents of a worthwhile human life (of human well-being or human flourishing) and the relations among human beings that are desirable for the sake of cultivating the well-being of everyone; they also concern (with special pertinence for some of the groups who wish to use agroecology in "developing" countries) the relations of human beings with the environment (and the numerous beings that compose it), as well as with other beings that may be recognized in a culture's worldview (including God), insofar as they are connected with human well-being. Like other kinds of values (e.g., social and cognitive values—section 3.1), ethical values are articulated in words enabling them to be objects of discussion, interpretation, and argument, and to constitute the basis for ethical norms, judgments of obligation, and legitimacy, which play roles both as criteria for choice among possible courses of action and as standards for appraisal of behavior, institutions, and social structures. They are also manifested in behavior, embodied in institutions, practices, policies, and social structures. An ethical value is manifested in one's behavior, for example, when it functions as a fundamental factor in explaining the goals that one adopts and the commitments one enters into. It is embodied in an institution when there are roles readily available in the institution for manifesting the value. There are gaps (discrepancies, tensions, disconnects) between the ethical values that one articulates as one's own and those manifested in one's life.[6] Such gaps have various sources: first, from the fact that our aspirations can and often should go beyond current realities; secondly, from social pressures to conform or lack of self-understanding; and thirdly, from the fact that we must live our lives in association with others—human action is almost always interaction with other humans—so that to manifest a value may depend on the reciprocity of others and this, in turn, can often only be counted on where there are institutions or movements available in which such reciprocity is nurtured.

To hold an ethical value authentically is: (i) to desire to live a life in which the value becomes progressively more fully manifested in one's behavior; (ii) to believe that a worthwhile life is marked by the high manifestation of this value; and (iii), elaborating (i), to commit oneself to a life trajectory in which the gap between its articulation and manifestation is progressively narrowed—and this will generally mean participating in institutions that manifest related social values. These three conditions ensure that authentic ethical commitments cannot remain static or complacent. I have already indicated that narrowing the gap can require available social conditions. If they are not currently available, something must adjust; either one adjusts (or resigns oneself) to what is currently achievable within prevailing social structures and their current trajectories, or one enters into a movement aiming for social transformation. Some values can only be held coherently if one lives within a movement that embodies those values in anticipation of a fuller transformation of society—think of values like solidarity with the poor, the fuller implementation of economic/social/cultural rights, or environmental sustainability.

Reflection on ethical values is deformed when the gap is ignored, with the consequence that ethical values are reduced either to those that are articulated or to those that are manifested and embodied highly (Lacey and Schwartz 1996; Lacey 2003d). When reduced to what is affirmed in words, ethics becomes simply a discourse of commendation and condemnation, a sort of "ethical fundamentalism," separated from one's desires, one's aspirations, one's sense of what is genuinely possible, and one's attempts to define the trajectory of one's life, and it is without links with practices of transformation—ethics then functions as a "moral club" to demean opponents and as a discourse to reinforce the self-righteousness of those who affirm favored values and norms. When reduced to what is highly manifested and embodied, ethics becomes identified with the hegemonic moral vision of a society and undermines any aspirations for social transformation. In neither case does appeal to ethics enable there to be constructive dialogue between those who hold different moral visions.

Holding particular ethical values rests upon certain presuppositions. I emphasize those about what is possible—is it possible, given prevailing social conditions and the actual aspirations of people, to manifest and embody a value (e.g., solidarity with the poor) more fully (chapter 11)? And also those about human nature—about what constitutes the distinctiveness of human beings and the dimensions of human well-being.[7] Largely because of these presuppositions, ethical values are essentially contestable; but also, again because of them, any contestation can potentially be (in important respects) an object of rational discussion and in many instances of empirical investigation (chapter 11). Sometimes, as we will see in section 10.5, beliefs in such presuppositions, which are not supported by empirical evidence, are the source of failure to perceive that certain gaps between articulation and manifestation are unbridgeable. There may be some ethical values that are universally shared; if so, that should be seen as reflecting a consensus arrived at in the course of dialogue conducted over the centuries. Authentic ethical discourse, however, does presuppose that all persons are, in principle, participants in the discourse, and that the (long-term) objective of well-being for everyone (to the extent that it is possible) frames the discourse—thus, inequalities of power are potential sources of deformation of ethical discourse.[8]

"Lack of Ethics"

I mentioned above charges that have been made about "lack of ethics" in government policy-setting and corporate behavior in connection with TGs. Often, when such charges are made, the crucial distinction is overlooked between those who hold a different (and incompatible) ethical vision to one's own, and those who act and enact programs in total disregard for the interests and rights of others (especially of the poor). Only the latter are properly labeled "unethical." But the two should not be identified, even when "self-interest," or the prioritizing of individualist values, is at the heart of the ethical vision that one opposes, for in that moral vision self-interest is articulated as part of a conception of freedom (that highlights individual choice) balanced by recognition of the value of civil/political rights and the constraints of a democratic society. Such an ethical vision is widely articulated today;[9] and it should not be identified with what may sometimes be actually manifested where it is articulated, that is, with unabashed self-interest and the cynicism that come with it, accompanied by corruption, violence, and ignoring of the law,[10] and even by using the agents of the law-enforcing apparatus of the state to serve that self-interest; even though it is true that major and serious gaps between articulation and manifestation of this ethical vision have enabled inordinate harm to be legitimated in recent years. Today, throughout the world, with the consolidation of neoliberal institutions and policies, the lapses from the individualist ethical vision may be more in evidence (certainly they are under more public scrutiny) than its authentic implementations. Even if this is the case, maintaining the distinction is of utmost importance; if there is a lack of ethics, we still need to ask: what ethical vision should be adopted in personal and public life?

The proponents of using TGs in agriculture tend to dismiss out of hand the charge that they are unethical. They acknowledge risks, of course, while maintaining, in accord with P_3, that demonstrated risks can be adequately managed and regulated. Confident of the products and promise of science and emboldened by its past successes, they are unmoved by the appeals to proceed with special caution. Moreover, despite the fact that (for the critics) TGs have become a symbol of the abuses of big business and its willingness to subordinate human life and democratic values to the interests of capital, they do not concede the ethical high ground to them. On the contrary, they counter that using TGs permits high productivity combined with friendliness towards the environment and, as already mentioned, they propose that it is necessary to feed the world (P_4). From this perspective any risks occasioned by the use of TGs fade into insignificance compared with the consequences of not developing and using them. In reality, they insist, it is their critics who lack proper ethical concern (McGloughlin 1999; Rauch 2003). No doubt, one can find critics who do lack proper ethical concern: to oppose TGs on the ground of risks (C_3) while ignoring the matter of alternatives (C_4) would be unethical. (The proverbial environmentalist who acts to preserve the environment in disregard for social justice for the poor no doubt can be found somewhere!) I also do not doubt that the corporations that have developed TGs have (sometimes or even often) lapsed into unabashed self-interest and permitted corporate interests to override not only genuine democratic concerns but also adequate assessment of risks. Nevertheless, these corporations think of themselves as bearers of a progressive ethical vision and (within its perspective) as subject to appropriate ethical constraints. For them, I repeat, it is their opponents, not they, who lack proper ethical concern. They maintain that, within *any* value-outlook worthy of serious consideration, TGs should be considered objects of high ethical and social value; not only is the development and implementation of TGs legitimate, it is virtually obligatory.[11]

Whatever one may finally make of all this, conflict between two opposed ethical visions is there at the heart of the controversy about TGs. The P- and C-sides and the different approaches to farming which they (respectively) highlight, embody fundamentally different value-outlooks.

Value-Outlook of the Proponents

For the P-side, the development of TGs derives from an ethical vision that, throughout the world today, is widely regarded as inevitable, even among some people who do not find it very attractive. It is an ethical vision that is deep in the "common sense" of modernity, one that puts great hope in techno-scientific "breakthroughs." The modern valuation of control (section 1.1) is at its core. Although it is deeply embodied in capitalist institutions and today neoliberalism is its foremost bearer, its appeal far transcends them, in part because of the special association that it claims to have with science. Especially when we think about medicine, there are few who lack a measure of sympathy for this vision. The modern valuation of control is integral to predominant conceptions of "modernization" and "development"; and part of the legitimation sometimes offered of neoliberalism is that it furthers the embodiment of these values in society "globally."[12] By way of a summary statement, neoliberalism may be identified with programs and policies that support, for example, private control of the economy, deregulation, removing restrictions on capital flow across borders, production for profit in the global market, reduced role and responsibilities for government (especially concerning social issues, though not concerning fiscal and monetary policies); and that are shaped by organizations (e.g., IMF, WTO, NAFTA, and the proposed FTAA) aiming to strengthen the international free market. The values held by neoliberal institutions may be considered elaborations of the more vague "modernization," "development" (*SVF*: chapter 8), and "globalization" that are integral to current global-market institutions and practices. They include (in addition to the modern valuation of control—see below):

- individualism (the primacy of holding egoist values)
- private property, private initiatives (and expanding the domains in which they can be taken) and profits
- expanded commodification (the reduction of value to economic value)
- individual liberty and economic efficiency
- the prerogatives of wealth with respect to socioeconomic innovations
- formal electoral democracy and the separation of powers
- the primacy of civil/political rights
- recognition of intellectual property rights

Value-Outlook of the Critics

The legitimating claim of neoliberalism has been that, following implementation of its policies, a rising standard of living will gradually spread throughout a country's population. According to the C-side, however, the reality has tended to be a growing of the gap between the handful of the rich and the vast majority of the poor (even if, in some "developing" countries numerically more people have become well-off), increased impoverishment and social disruption, the selling off of national patrimonies to foreign owners, decreased social services and educational opportunities for the poor, the devastation of hope and the accompanying rise of criminal and terrorist violence, the exposure of the poor periodically to especially intense sufferings induced by fluctuations of the global economy, and increasing cultural homogenization (Lacey 1997).

This assessment is informed by its competing values: prioritizing environmental sustainability, maintenance of biodiversity, caution in the face of risks to health (thus it contests the modern valuation of control), and (what have been called within the World Social Forum (WSF)—see appendix) the values of "popular participation." These include the following, listed so that, item by item, their contrast with the values of neoliberalism is apparent:

- solidarity and compassion (in balance with autonomy) rather than individualism
- social goods (and the role of government in furthering them in areas such as education, health care, and land reform) balancing private property and profits; the well-being of all persons rather than the primacy of the market and property
- strengthening a plurality and diversity of values rather than expanded commodification, and protecting the "common assets of humanity" from commodification
- human emancipation (liberation) as encompassing and qualifying individual liberty and economic efficiency
- the rights of the poor (and the importance of their initiatives) and the primacy of life (and the enhanced agency of everyone) prioritized over the interests of capital
- participatory as encompassing formal democracy, where democracy has an international dimension
- civil/political rights in dialectical relation with social/economic/cultural rights, usually taken to include commitment to nonviolence (and emphasis on dialogue and tolerance) to the extent that it does not involve the toleration of injustice
- intellectual property rights subordinate to social/economic/cultural rights

Truthfulness, Impartiality, the Modern Valuation of Control

Truthfulness may be considered an important value for both the proponents and the critics, and this includes commitment to impartiality. The P-side, since it upholds the modern valuation of control (and thus privileges materialist strategies in scientific inquiry), can easily also come to value deference to the authority of techno-scientific "experts." The C-side, since (see below) it contests the modern valuation of control, is led to interpret truthfulness in the context of conducting research under a plurality of strategies, including those linked with culturally specific and traditional forms of knowledge. It aspires to gain comprehensive understanding of the place of our lives in the world and to identify the liberating possibilities hidden within the predominant order, and it does not identify what is possible with the principal tendencies of this order. Moreover, since it is committed to impartiality, it values being prepared to submit to criticism and investigation the legitimating presuppositions of one's practices, rather than to place them among "certitudes" that are seen to be beyond investigation, as can effectively happen when one is not attuned to alternatives (chapter 11).

6.4 EFFICACY AND LEGITIMACY

In the light of my account of the protagonists and their respective values, the controversy may seem to defy rational resolution. It may even appear that opposition to TGs is (as the P-side says) simply resistance to both techno-science and economic progress, resistance to the trajectory of the contemporary world. As pointed out in section 6.1, the efficacy of TG technology is not seriously in doubt; among other things, there are TG crops that are resistant to commonly used herbicides and toxic to certain kinds of insects. These technologies work. Some might think that the claims to efficacy have been exaggerated, or that efficacy does not translate into efficiency, or that the long-term efficacy and sustainability of TG technology is not settled, or that the stability of TGs beyond a few generations is in doubt, or that many of the promises for the future are mere hype—all important matters—but it is pretty well settled knowledge that the TG technology currently used is for the most part efficacious. Moreover, the farmers who use them, the corporations that produce them, and the global trade organizations that pressure nations to use them clearly consider TGs beneficial; and, for example, other groups (see appendix) are pushing for developments of TGs that, they maintain, can serve the interests of poor farmers and nations. What else could ground denying the legitimacy of using them other than resistance to techno-science and economic progress?

The answer is obvious: serious concerns about risks and valuing alternative approaches more highly (C_3 and C_4) and about the value of the alleged benefits (C_2). Yet, often the P-side treats this answer simply as a cover for resistance to techno-science and economic progress (see section 10.2). It treats P_3 and P_4 as virtual certitudes. P_4 rarely gets explicit mention, and so arguments are normally not mounted against C_4; appeal to P_1 is apparently taken as sufficient to endorse P_4 (again, see section 10.2). P_3 is regularly mentioned, typically accompanied by the assertion that its scientific credentials are beyond reasonable doubt, having successfully met all the scientifically credible challenges that have been made. The C-side questions the adequacy of the scientific evidence for P_3 (section 9.4; section 9.5). The questioning is based in part on the fact that risk assessment is necessarily comparative, and so its adequacy depends on comparisons with appropriate alternatives. If conventional high-intensity farming is the only point of comparison, the evidence for P_3 may indeed be compelling. But the C-side denies that this is the relevant point of comparison, for it endorses C_4 and, for it, the forms of farming (e.g., agroecology) that they favor are considered to be alternatives to both TG-oriented and high-intensity farming.

There cannot be compelling scientific evidence for P_3 unless there is compelling evidence for P_4, and that evidence cannot come from the same research that addresses the efficacy and possibilities of TG technology. Testing P_4 requires also appraising C_4, and that cannot be done in research conducted virtually exclusively under materialist strategies, such as the research conducted under biotechnological strategies that has led to the developments of TGs. To appraise the possibilities that might emerge in agroecology, one must adopt strategies (not reducible to materialist ones) for research, within which objects (e.g., seeds and crops) are not dissociated from their ecological, human, and social dimensions. Testing P_4, therefore, requires that scientific research not be limited in the way deemed exemplary in P_1 (i.e., to research conducted under materialist strategies). Otherwise the potential of agroecology cannot be empirically appraised. In order to test P_4, research needs to be conducted under a variety of strategies, at least agroecological strategies (which do not dissociate phenomena from their sociocultural context) as well as those, like those adopted in biotechnological research, under which we gain knowledge of the underlying structure and law of phenomena (which do).

Furthermore, if research were to show that there are no genuine alternatives and thus to provide compelling evidence for P_4, then greater risks should be tolerated (given general agreement that current conventional high-intensity forms of farming are unsustainable) in developments and implementations of TGs, than if there were such alternatives. Not to take the risks, if P_4 were empirically well confirmed, would be to risk creating serious food problems in the future. But P_4 cannot be empirically well confirmed unless the evidence offered by the critics for endorsing C_4 is rebutted, and it has not been (see section 10.1; section 10.3). Strong endorsements of P_3 and P_4, at present, are not driven by sound appeals to scientific evidence, but presupposing P_1 makes it difficult to recognize that this is so, especially since it tends to be taken for granted unreflectively.

The Modern Valuation of Control

Here we see the influence of the modern valuation of control and its presuppositions ((a)–(e) in section 1.1) on the argument of the P-side. Endorsement of P_1 makes sense in the light of its instantiating Presuppositions (a) (techno-scientific innovation serves human well-being) and (b) (ability to find techno-scientific solutions to major problems),[13] and the mutually reinforcing relations between holding the modern valuation of control and adopting materialist strategies. In addition, the hasty endorsement of P_3 fits with readily making the move from efficacy to legitimacy that is encouraged by the fourth component of the modern valuation of control. (The values that may be manifested in social arrangements are, to a significant extent, subordinate to the value of implementing novel techno-scientific advances, which have prima facie legitimations, so that a measure of social disruption may be tolerated for their sake, and whose side effects [risks] may be addressed largely as "second thoughts.") And C_4 is ruled out of contention by Presupposition (d) (there are no significant possibilities for value-outlooks which do not contain the modern valuation of control to be actualized in the foreseeable future), since the alternatives it cites are linked with the values of popular participation or other values that contest the modern valuation of control (see section 10.6; section 11).

Items 3 (risks) and 4 (alternatives) involve disagreements about matters of fact, and so empirical (scientific) inquiry is pertinent to settling them. But we find (chapter 9 and chapter 10) that the factual disagreements are intertwined with further disagreements about (i) the strategies that should be adopted in relevant research, (ii) the standards of evidence to which claims on these matters should be held, and (iii) what approaches to agriculture are worth exploring for their productive potential. Disagreements concerning (i) and (ii), which are influenced by those at (iii), can be expected to lead to disagreements about the adequacy of the evidence supporting the propositions in each item, in particular about whether the proposition is properly *endorsed* (section 3.4), about whether or not it has sufficient degree of confirmation to legitimate (or not) actions that are informed by it. Social (ethical) value judgments are overtly involved at (iii). They also are involved at (i) and (ii). Commitment to the modern valuation of control explains why the P-side can endorse P_4 while effectively ignoring (and not rebutting) evidence proposed for C_4; given the interests served by this valuation, too much would be at risk if C_4 were acted upon (and no additional investigation would affect that judgment). It also explains the kinds of research deployed in risk assessments and the standards of evidence that are expected to be met, so that the burden of proof is firmly placed on the critics, and their claims are expected to be held to very high standards of evidence (section 9.4). Similarly, commitment to the values of popular participation (or other value-outlook contesting the modern valuation of control) lies behind strong endorsements of

C_3 and C_4, and the expectation that endorsements of P_3 and P_4 should be held to very high standards of evidence (which at present they are deemed not to meet).

Provisional Endorsement

None of the propositions in items 3 (risks) and 4 (alternatives) can at present be accepted in accordance with impartiality (and so the authority of science does not back endorsement of any of them). But we cannot simply wait for further investigation on them, for we cannot defer taking a stance (at least a provisional one) on the legitimacy of TG-oriented farming. That question is on the public agenda—put there not by scientists, philosophers, or movements of small farmers, but by agribusiness corporations and the governments whose policies support them—and it will not go away by itself. Arguments for (or against) the legitimacy of using TGs draw upon endorsements of some or other of these propositions. I use "endorsement" to make clear the contrast with sound acceptance in accordance with impartiality. Endorsements reflect social (and ethical) value judgments in interplay with cognitive value judgments (see section 9.6). That does not cast into doubt, however, that here we are dealing with disagreements about matters of fact, and not just with the present state of the evidence and the stances towards action that it may allow. Acceptance, in accordance with impartiality, remains the scientific aspiration, even if endorsement is the best we can do at the moment.

Could not all parties recognize, then, that their endorsements are provisional, pending the conduct of further appropriate research? Taking such a position may be the key (if there is one) to resolving the controversy about TGs. (The outcome of conducting such research might not be to legitimate either the P- or C-side, but a range of more moderate options.) The aim of science and upholding the value of neutrality would lead to taking it.

The C-side could live comfortably with it; the fuller manifestation of its values depends on the development of its favored alternatives (in agriculture as well as in other domains), so that its interests are likely to be served by further research on the productive potential and sustainability of alternative forms of farming. If this research were properly conducted there is the possibility that it would lead to the decisive refutation of C_4 (and also C_3). (Given the C-side's assessments of the promise of the alternatives—see section 10.1—its proponents would consider this an unlikely outcome.) Were that to happen, the argument of the C-side would collapse, and that would put into serious doubt the possibility of fuller manifestation of the values of popular participation and provide some evidence for Presupposition (d) of the modern valuation of control (no possibilities not linked with the modern valuation of control). For those who hold the values of popular participation this would, no doubt, be thought a devastating outcome, but, I submit, it is a mark of holding values rationally that those who hold them be prepared to put their presuppositions (where possible) to empirical tests, and to reconsider their value commitments if their presuppositions cannot be sustained.

Appropriate research, if it is genuinely empirical, must be conducted under strategies chosen so that, antecedent to conducting it, it is not effectively determined whether it will tilt in favor of P_3 and P_4 or C_3 and C_4. That means it must be conducted under a pluralism of strategies, and so C_1 will need to be (at least provisionally) presupposed. As just indicated, presupposing C_1 does not guarantee that C_3 and C_4 will actually gain greater empirical support. It does, however, underlie conditions that are needed for appropriate empirical investigation of items 3 and 4, and does not preclude—antecedent to carrying out the relevant research—that P_3 and P_4 might gain strong empirical backing. Moreover, P_3 and P_4 cannot gain strong empirical backing unless C_1 is presupposed.

Even so, many on the P-side would hesitate to agree that their endorsements are provisional.[14] After all, they want to legitimate the intensive and more widespread use of TG crops in present-day as well as future agriculture, whereas provisional endorsements of P_3 and P_4 would underscore the need for further research concerning them, and thus undercut that there is a scientific basis that lends urgency to utilizing them now on a grand scale. There is strong interest among certain groups to push ahead rapidly with using TGs. Agribusiness corporations are interested to make profits from the enormous capital investment that they have made in developing TGs and to expand their dominance into new sectors of agriculture. Some governments see these developments as crucial to their trade policies; and some scientists have benefited from gaining patents to aspects of TG-technology, and they see research connected with TGs as an important inducement to continued financial support of scientific research and a source of jobs for those with training in biotechnology (see section 10.6). These interests are linked with the modern valuation of control, and thus they contribute to reinforcing the presupposing of P_1. We have seen, however, that where P_1 is presupposed, conditions are not made available for empirically testing P_3 and P_4 with appropriate rigor; and its being presupposed is explicable principally in view of its relationship with the modern valuation of control (and the values linked with it). That is, the conditions are not available for empirically testing P_3 and P_4 with appropriate rigor because certain social values are held. Then, where P_3 and P_4 are taken to be accepted (as effectively needing no further empirical testing), impartiality is violated, for these social values are then effectively playing a role alongside the cognitive values at the moment of accepting scientific knowledge.

Research Conducted under a Plurality of Strategies

Social values by themselves, even a set of values as widely held as the modern valuation of control, independently of which propositions in items 3 (risks) and 4 (alternatives) are true, do not suffice to legitimate either the P- or the C-positions. Something else is needed to break the impasse that marks current controversy about TGs, but currently established scientific knowledge is unable to do so. I suggested above that the key to breaking the impasse (if there is one) lies in recognizing that current endorsements (concerning items 3 and 4) are provisional and that there is a need for further research concerning them. How can that be so, when my own analysis suggests that value judgments (fundamentally opposed ones) cannot be avoided in the contexts of appraising P_3 and P_4 and their respective contraries? Well, my analysis also shows that there are dialectical relations between (i) the outcomes of empirical research on risks and alternatives, (ii) the evidential standards brought to bear in them, and (iii) the values that are implicated in one's sense of what is a benefit and what may be at risk. While value judgments cannot be derived from scientific results, scientific results sometimes can provide compelling evidence that certain values cannot be more fully embodied, and that can reasonably motivate rethinking of value judgments (cf. the discussion of conducting inquiry on C_4 in the previous subsection). Conversely, value judgments can motivate adopting some strategies rather than others.

The value disputes in play in the disputes about TGs do not reduce to differences of opinion that could be definitively settled by empirical investigation. Nevertheless, the propositions in items 3 and 4 are open to empirical investigation, which could recast the controversies. Minimally, unless the productive potential of alternatives, such as agroecology, is confirmed by well-supported research conducted under agroecological (and related) strategies, the case for C_4 would collapse, and that for P_4 (and thence P_3) would be strengthened. Since the relevant research requires the use of strategies in addition to materialist ones, it cannot be conducted where the view of what is exemplary scientific knowledge, expressed in P_1, is presupposed.

The need to conduct research, aiming to appraise P_3 and P_4, under a multiplicity of strategies is not always recognized; indeed P_1 is often simply taken for granted. That is why I said at the outset of this chapter that the controversies about TGs raise questions about the nature and proper mode of conduct of scientific research. Here ethics and philosophy of science are deeply intertwined. Those who adopt the values of the C-side have an interest in developing research under agroecological strategies. I hope that it is now clear that it is not only their values that lend interest to such research. Science itself has an interest, for it is fundamental to the aims of science that no claim about the world—including those of the form "there is no . . . "—be accepted under the authority of science, unless it has passed the rigors of empirical testing. Science has this interest even if the conditions for satisfying it (e.g., support for furthering agroecology) are in tension with the widespread acceptance of the modern valuation of control and the thrust of neoliberal policies and projects, and even if this thrust is considered to be effectively irresistible (section 10.6; chapter 11). Together, the aims and methods of science, criteria used to evaluate scientific knowledge, and the record of research conducted under strategies (like agroecological ones) that are not reducible to materialist ones, do not provide a compelling argument for P_1, but rather point towards C_1.[15] Recognizing this, of course, implies that (at best) only provisional legitimacy is available at present for the development and utilization of TGs, but it is consistent with their gaining strong legitimacy in the long run, depending on how research (that provisionally presupposes C_1) turns out.

One by one in the next four chapters, I will discuss in much greater detail each of the four pairs of propositions stated in section 6.2. Inevitably there will be some overlap and considerable cross-referencing among these chapters, for all the propositions are deeply intertwined in playing their roles in the respective arguments. My discussion will be informed by three "morals" that I draw from the analysis of this chapter: (1) Clear thinking about the legitimacy of using TGs requires grasping the crucial interplay of values and scientific research. (2) There is an urgent need for further scientific research (which does not prejudge whether or not there are legitimate roles for TGs in ongoing and future agriculture) on the productive potential of alternative forms of agriculture; for the risks that may be occasioned by using TGs may legitimately be taken only if less risky or more valuable alternatives are not available. More generally, well-grounded judgments about the legitimacy of using TGs should reflect answers to the question: How can we produce crops so that all the people in the region of production will gain access to a well-balanced diet in a context that enhances local well-being, nourishes biodiversity, sustains the environment, and supports social justice? (3) While scientific input is essential for appraising the legitimacy of using TGs, the issues involved go far beyond the specific competence of molecular biologists and genetic engineers. Consequently, public policy on these matters should be determined, not by commissions of scientists, but by bodies that involve the participation of representatives, including both molecular biologists and researchers in agroecology, reflecting the diversity of value outlooks held throughout society.

NOTES

1. Ho's article contains nice short accounts of the basics of transgenics as well as of her strong criticisms of their uses.

2. "TGs" (transgenics) refers to TG (transgenic) plants. I will also refer to TG seeds, TG organisms, TG crops, TG technology, TG products, and TG-oriented farming. TG organisms are often called "genetically modified organisms" (GMOs). Details about the techniques of genetic engineering, the biotechnological and molecular biological knowledge that informs

them, and the variety of TGs that have been produced and those under development, can be found in any up-to-date textbook that deals with agricultural biotechnology (e.g. Alcamo 1996).

3. I use this term having in mind Wittgenstein (1958: part 1, paragraph 122) on "perspicuous representation."

4. In part I, on several occasions I noted that within mainstream scientific circles (where the P-side locates itself), what I call "research conducted under materialist strategies" is usually simply considered to be "scientific research." Thus, we would expect that spokespersons of the P-side would omit the clause, "that is, . . .", from their own formulations. I include it in my formulation because I do not want to limit what counts as exemplary scientific knowledge in this way for that would, effectively by definition, rule the C-side out of contention. It would also, as we will see, put investigation of P_4/C_4 outside of the realm of scientific investigation.

5. Unless C_1 is endorsed, it does not seem to be possible to produce an interpretation of the controversy which permits portrayals that each side can acknowledge as containing a fair representation of its position. One can endorse C_1 (now) and understand why others endorse P_1 (links with the modern valuation of control). But, if P_1 is taken as a condition on adequate interpretation, the C-side would not be represented in a way that it acknowledges as accurate, for under that condition it would be represented as "unscientific," or even "anti-scientific."

6. Analogous "gaps" exist in connection with all kinds of values. I identified one in the practices of scientists in the discussion of "professionalism" in section 2.2.

7. Additional presuppositions can be located by extrapolating my analysis of social and cognitive values in section 3.2 to ethical values. Note that (in section 1.1) I have portrayed the modern valuation of control (a set of social values) as drawing upon presuppositions about what is possible, about social ideals, and a conception of nature. One important kind of question about what is possible concerns whether two articulated values (e.g., the modern valuation of control and respect for human rights) can be highly manifested in the same institutions.

8. I am of the opinion that adopting what Latin American liberation theology—an important influence upon some of the movements that are on the C-side (section 6; appendix; section 11.2)—calls "the preferential option for the poor" contributes to strengthening the authenticity of ethical discourse (Lacey 1985; 2003c).

9. Consider: "America is successful and wealthy because of its values, not despite them. It is prosperous because of the way it respects freedom, individualism and women's rights, and the way it nurtures creativity and experimentation. These values are our inexhaustible oil wells" (Friedman 2002).

10. Compare: "Enron's failure was a failure of particular people and institutions, but it was above all part of a general failure to maintain ethical standards that are, in my view, fundamental to the American economic system. Without respect for those standards, popular capitalism cannot survive" (Rohatyn 2002).

11. On this point, there are varying emphases; see, for example: Borlaug (2000); Human Development Report (2001); Nuffield Council on Bioethics(1999); Persey and Lantin (2000); Potrykus (2001); Serageldin (1999); Specter (2000).

12. "Globalization" is a contested term. On the one hand, it refers to the fact of increased interaction among peoples from all parts of the world and to the possibilities for cooperation that this occasions. On the other hand, it refers to the structuring of this interaction within the neoliberal order. The World Social Forum (see appendix) values the former but not the latter. Note also that the term "neoliberalism" is not much used in political discourse in the United States, where most of what it refers to falls under such labels as "the Washington consensus,"

"free trade," "democracy," and "freedom." I prefer to accompany the critics in using the term "neoliberalism," in part because I do not want to concede the terms "democracy" and "freedom" to the uses it has in this discourse. What is called "neoconservatism" in the United States, while admitting of variety that in some cases opposes "globalization," tends to build a conservative social agenda on top of the neoliberal economic policies.

13. Note that the presuppositions of the modern valuation of control themselves cannot be investigated adequately in research conducted under materialist strategies (chapter 11). Ironically, if I am right, the predominant role granted to materialist strategies in modern science is rationalized in view of commitments to the modern valuation of control, but the latter rests on presuppositions that themselves cannot be adequately investigated under materialist strategies.

14. I have in mind agribusiness corporations and their governmental and scientific allies who are lobbying hard for the accelerated utilization of TGs. Scientific bodies which support research on TGs and anticipate that they will play a positive role in the farming of the future may be more likely to take a provisional stance.

15. The controversies about TGs are marked by lots of demagoguery. I am trying to cut through this to what I think is really at stake. I think that commitment to the modern valuation of control best explains why P_1 tends to be taken for granted; it is also convenient cover for the bullying tactics of those with strong interests in not wanting further testing of P_3 and P_4 to be conducted. I have claimed that the value of neutrality, widely acclaimed in scientific circles, points towards presupposing C_1, but others see it as cover for wishful thinking that they assert to lie behind C_4.

Chapter Seven

Strategies for Research in Agricultural Science

P_1 Developments of transgenics are informed in an exemplary way by scientific knowledge, that is, they are informed by knowledge gained in research conducted under appropriate versions (biotechnological) of materialist strategies; they are instances of techno-scientific developments, which are the principal sources of improvements of agricultural practices and (more generally) meeting human needs.

C_1 The kind of knowledge gained under materialist strategies is incomplete and cannot encompass the possibilities of, for example, sustainable agroecosystems and the possible effects of uses of transgenics on the environment, people, and social arrangements; it is necessary to adopt other strategies in order to investigate these matters.

Questions about scientific knowledge must be central to the discussion of TGs. TGs are products of scientific knowledge gained by conducting research under versions of materialist (biotechnological/molecular biological) strategies, and questions about their efficacy are, in principle, settled by reference to this knowledge. Scientific knowledge is needed also to address responsibly questions about risks and alternative approaches to agriculture, and thus it should be central to discussions of the legitimacy of using TGs. When addressing legitimacy, not only questions of scientific knowledge need to be addressed, but also questions of scientific methodology. What strategies need to be adopted in order to investigate risks and alternatives adequately? I have already mentioned (section 5.4; also appendix: 1) that the P-side often colors the C-side with the "anti-science" brush, thereby insinuating that the latter's case is purely ideological. I have countered that at the heart of the dispute are both value ("ideological") and scientific matters. The latter include making decisions about what strategies to adopt in research, and, in turn, they have implications about what scientific knowledge becomes available for application in farming practices. I have encapsulated the matter in dispute here in the contrasting propositions P_1 and C_1. Throughout part I, I repeatedly made arguments for the possibility of conducting scientific research under a plurality of strategies, and also for the necessity of doing so for the sake not only of the interests of those who contest the modern valuation of control, but also for the further manifestation in scientific practices of the value of neutrality. In this chapter, I will both explore the appeal of P_1 and elaborate arguments of part I that there is a solid prima facie case provisionally to endorse C_1. (As pointed out in section 6.4, endorsing C_1 leaves open that, after conducting relevant research, the evidence may support endorsing P_{2-4}.) The C-side is not "anti-science"; but it does reject reducing what counts as scientific research to that which is conducted under materialist strategies.

7.1 SEEDS AND THE KNOWLEDGE THEY EMBODY

In modern consciousness the achievements and promise of science loom large. So too do the expanded human powers to exercise control (in technology) that have been unleashed by techno-scientific developments. While for some, science and new technologies induce fear and apprehension, for the most part in the contemporary world their value has been deeply internalized. Thus widespread legitimacy has been accorded to research and developments into novel technological possibilities, and it tends to be taken for granted—though not without opposition—that the future will, even must, be shaped largely in response to them. The modern valuation of control (section 1.1) has been widely embraced and its presuppositions widely accepted. TGs and other biotechnological "breakthroughs" are among the latest and most visible successes of applying knowledge gained under materialist strategies in practices that express the modern valuation of control. To many on the P-side (chapter 6; appendix), insensitive to the potential pluralism of strategies and convinced that there are no alternative paths outside of those framed by the modern valuation of control (section 6.3), TGs clearly are the way of the future in agriculture, and they also testify to the remarkable ingenuity and providence of science. Their rhetoric often displays a sort of breathless and awestruck amazement at the explosive growth of knowledge of genes and genomes and of new "possibilities" that are promised by biotechnology.[1] A widely used textbook is entitled *DNA Technology: The Awesome Skill* (Alcamo 1996). That sums it up nicely. To criticize biotechnology (and its TG products) seems to be on the verge of blasphemy, to be posed against the unfolding future and against science itself (see also section 10.6). The use of religious-like language is quite common in celebrating the alleged benefits of

TG seeds; for example, some of them are commonly referred to as "miracle seeds" and "grains of hope" (Nash 2000).[2]

The legitimation of the development and deployment of TGs is typically sought in the authority and prestige of science, and this is supposed to silence all critics.[3] To their developers, TGs embody scientific knowledge; they bear the imprint of science. They also bear the imprint of the political economy of "globalization," since developing them has been seen as both an objective of the neoliberal global economy and a means towards entrenching its structures. The twin imprints lend an aura of inevitability to the agricultural "revolution" promised with the advent of TGs: science has set the course; the global economy provides the structures for its effective implementation. Little wonder, then, that the growing of TG plants (corn, soybean, and others) has expanded explosively in the past few years. Nevertheless, as sketched in section 6.4, science (by itself) does not legitimate this explosion, and it poses no barrier to exploring alternative forms of agriculture that are informed in part by knowledge gained under strategies that are not reducible to materialist ones.

Granting Intellectual Property Rights to TG Seeds and TG Technology

I said that, to their developers, TG seeds embody scientific knowledge. More accurately, they embody scientific knowledge gained under materialist strategies. Among those who hold the modern valuation of control, that is the key to the social value that these seeds are deemed to have and (in part) the key to the legitimation of making them bearers of IPR (intellectual property rights). Among others, it is the key to contesting their social value and opening the door to the importance of conducting research under agroecological (and related) strategies. I will dwell a little on the issue of granting IPR to TGs, for it will serve to highlight the mutually reinforcing relations between biological research conducted under materialist strategies and the modern valuation of control (and the values of property and the market that, in turn, reinforce it).

TG seeds are engineered by modifying the genomes of already available seeds. These may be farmer-selected (FS) seeds, many of which (as Altieri points out; see section 5.4) are the products of traditional local knowledge that reflect sound agroecological understanding. Or they may be seeds, originally derived from FS seeds, which have been selected by breeders for use in conventional farming (or for producing earlier varieties of TGs). The very existence of TGs requires the prior development of FS seeds (Kloppenburg 1988). Yet IPR protections may be granted to TGs but not to FS seeds. Lacking these protections, FS seeds are considered to belong to the common patrimony of humankind, and they may legally (under influential prevailing laws and international agreements) be appropriated at will without consultation with or compensation for the farmers (or their descendents) who selected them (Kloppenburg 1987); and seed banks (mainly containing germplasm from poor tropical countries where the world's centers of biodiversity are heavily concentrated) have been established under the auspices of international bodies to facilitate the development of new varieties of seeds. When FS seeds are so appropriated, critics speak of "biopiracy" and diagnose inequity. As they see it, the developer of TG seeds freely appropriates FS seeds, but the farmer does not have free access to the TG seeds. Not only agribusiness (through its research scientists) but also generations of farmers contribute to the production of TG seeds, but, thanks to IPR, mainly agribusiness corporations and their clients profit. Any such profits presuppose the free appropriation of FS seeds—and of the knowledge (traditional, practical farming, professional breeders', agroecological) embodied in them. TG seeds embody this knowledge as well as knowledge gained under materialist strategies; indeed it is by far the greater part of the knowledge embodied in TG seeds, since the genetically engineered modifications involve only small segments of the genomes. Moreover, the conditions in which profits are gained tend to facilitate the displacement of FS by TG seeds.[4] Biopiracy thus involves not only appropriation of the farmers' knowledge that then becomes embodied in TG seeds (a sine qua non of the existence of the latter), but also, in the end, taking away from them the very use of these seeds, thus undermining the conditions of livelihood of many small-scale farmers, their families, and communities (Shiva 1997a; 2000a; 2000c). What the critics call "biopiracy" and the regime of IPR are deeply interconnected. The development and deployment of TG seeds has depended on them both.

What are the differences between TG and FS seeds that can make sense of the fact that the former, but not the latter, can be granted the protections of IPR? One alleged difference is that TG but not FS seeds embody scientific knowledge. (Remember, I have maintained that it is more accurate to say that they embody scientific knowledge gained under materialist strategies.) In virtue of this they, but not FS seeds, may satisfy the standard criteria for gaining a patent—novelty, inventiveness, utility/industrial application, and provision of sufficient instructions to meet the "sufficiency of disclosure" condition—and thus become intellectual property. Law courts and legislatures (in many countries) have upheld that this is so. That's all there is to it! TG but not FS seeds qualify for IPR, so that only sheer demagoguery and sentimentalism can lead to calling the free appropriation and eventual displacement of FS seeds "biopiracy." The prestige of science (understood as inquiry conducted under materialist strategies) is thus cast against using such a morally loaded term. Only property may be pirated, and FS seeds are not intellectual property. This point is usually made in conjunction with P_2 (great benefits) and P_3 (no serious unmanageable risks), and also with P_4

(no other way to feed the world), so that the granting of IPR to TG seeds becomes linked with a humanitarian obligation. These links are important for meeting the "utility" condition for award of a patent. But then gaining the patent is based not just on the kind of scientific knowledge embodied in TGs, but also on the confirmation of P_{2-4} (or the variants of P_4 discussed in section 10.6). That confirmation (if it is to be forthcoming—as sketched in section 6.4, and developed in chapter 8–chapter 10) requires research conducted under a plurality of strategies. The utility condition would not be met if C_{2-4} were confirmed, especially if it were shown that one of the risks incurred by using TGs is undermining the conditions for successful agroecology, which (according to C_4) has an important role in addressing the food needs of small-scale farmers in impoverished countries.

It is easy to miss the force of this point about the utility condition when it is said simply that TG seeds embody scientific knowledge. Putting things this way takes for granted that science consists only of research conducted under materialist strategies and so, by resonating with the modern valuation of control, it carries a presumption in favor of P_{2-4}. It suggests that TG but not FS seeds can be informed by scientific knowledge, that is, by knowledge with superior epistemic credentials. I have maintained, however, that this is not a difference between the two kinds of seeds. The difference comes from embodying knowledge gained under different strategies, materialist or agroecological, respectively. Since the knowledge that is embodied in FS seeds is gained under strategies that are also pertinent to testing P_{3-4}, its epistemic credentials cannot properly be ignored. The kind of scientific knowledge embodied in TG seeds cannot provide the rationale for granting IPR to them and not to other kinds of seeds. More plausible, I suggest, is the converse: knowledge gained under materialist strategies is privileged (taken to have greater social value and sometimes, mistakenly, greater epistemic value) because on application it can be readily embodied in products with market value, including some that may gain the protections of IPR; so that the prestige of materialist strategies and the commonplace narrowing of the meaning of "science" reflect not superior epistemic credentials, but the greater social value of their applications among those for whom relations of control over natural objects or the economic value of things are prioritized. IPR are granted to TG seeds because they have value in the light of the interests of business and the market. The protections of IPR are needed to defend these interests; without them, investments in research and development of TGs could not be reasonably assured of a good chance of reaping profits, so that, without them, developing TG technology would not be a high priority. Science is not the key factor. Rather the mutually reinforcing relations between research conducted under materialist strategies and the modern valuation of control (and the socioeconomic interests that both further it and that it serves) provide it.

What TG Seeds Are

TG seeds do embody scientific knowledge gained under materialist (biotechnological) strategies, and they would not exist but for developments of molecular biology, such as the discoveries of the DNA structure of genes and the techniques of recombinant biotechnology. They are organic entities whose genomes have been modified by TG technology; they are biological objects *and* authentic products of techno-science. They also (for the most part) are intellectual property, objects of value within certain socioeconomic institutions.

7.2 SEEDS AND THEIR SOCIOCULTURAL LOCATION

It is a truism that the methodology of a scientific investigation should be appropriate to the nature of the object under investigation. Methodology reflects ontology. That is why the creators of modern science in the seventeenth century spent so much time presenting arguments that their methodology (early versions of materialist strategies) was appropriate in the light of their accounts of the nature of the material world. By the same token, scientific investigations of crop seeds and plants should deploy strategies that are appropriate in view of what they are. I said (in the previous section) that TG seeds are bearers of scientific knowledge gained under materialist strategies, and (for the most part) they are intellectual property. They are also biological objects whose genomes and underlying biochemical properties differ only a little from those of FS seeds. They are social objects as well as biological objects. Of course, to be the kind of social object that they are, they must have the biological properties that they do have; but they have been engineered—informed by knowledge gained under biotechnological strategies—to have the biological properties they have so that they would be this kind of social object. That they are both biological and social objects is relevant to the appraisal of P_{2-4}. Risk assessment, for example, which only entertains possible risks that may arise in virtue of the biological properties of TGs and the engineering processes deployed in making them, overlooks that the properties they have in virtue of being social objects may also occasion risks (see section 9.5). Understanding TGs *fully*—why they were developed, why are used, what social value they have, the full range of risks that they may occasion, their limitations, their prospects—requires locating them socioculturally and investigating them as social, as well as biological, objects.[5] This is an instance of a more general point: *what crop seeds are* is partly a function of the sociocultural location of which they are constituents, and their value does not significantly transcend their specific location. I will argue that the following two questions cannot be separated: How are seeds (plants and crops) to be scientifically investigated? How is the knowledge obtained from such investigations, on application, to be evaluated? The answers, in turn, vary with the sociocultural location.

What Seeds Are

Crop seeds (and the plants that grow from them) are simultaneously many things—including:

- Biological entities: under appropriate conditions they will grow into mature plants from which, for example, grain will be harvested.
- Constituents of agroecosystems.
- Entities developed, produced, and used in the course of human practices.
- Objects of social value, objects with which human beings may bear socially constituted relations, and perhaps having economic, legal, cultural, aesthetic, cosmological, or religious significance.
- Objects for empirical investigation; thus, objects that embody knowledge.
  (a) As biological entities, they are subject to genetic, physiological, biochemical, cellular, developmental, etc. analyses;
  (b) as parts of ecological systems, to ecological analyses;
  (c) as products of human practices, to analyses of their roles and effects in the sociocultural location in which they are planted and their products distributed, processed, consumed, and put to other uses; and, more generally,
  (d) as objects of social value, to a variety of social scientific investigations.

The specific ways in which seeds are all of the above kinds of entities, and the specific possibilities that are open to them, vary systematically with the sociocultural location of farming. Seeds used in farming may be and traditionally have usually been biological entities that are reproduced simply as part of the crop harvested. As such, they are renewable regenerative resources that (conditional upon a measure of social stability and absence of catastrophes) may be integral parts of sustainable ecosystems that generate products that meet local needs while being compatible with local cultural values and social organization and that have been selected by numerous farmers over the course of centuries with methods informed by local knowledge (Shiva 1991; 1997a).[6] Traditionally such seeds have been considered to belong to the common patrimony of humankind, available to be shared as resources for replenishing and improving the seeds of fellow farmers. In contrast, seeds may be commodities: objects bought and sold on the market, "property" whose users may not be their owners, whose features and uses are integrally connected with the availability of other commodities (e.g., chemical inputs and machinery for cultivation and harvesting), and that sometimes can be patented and otherwise regulated in accord with IPR. Under these conditions, they are developed by professional breeders and scientists and produced largely by capital-intensive corporations. Then, they cannot be understood simply (and sometimes not at all) as part of the grain harvested, or as components of stable ecosystems, and certainly not as entities to be freely shared with fellow farmers.

Increasingly throughout recent history seeds have been transformed from being predominantly regenerative resources into commodities. The transformation, whose mechanisms have been well described by others (Kloppenburg 1988; Lewontin 1998; Berlan 2001; Shiva 1991; 1997a), was initiated with the introduction of "high-intensity models" into agriculture, models based on mechanization and the use of extensive chemical inputs (fertilizers, pesticides, herbicides, etc.); and then further developed by the planting of monocultures, of hybrid seeds that do not reproduce themselves reliably and so must be bought regularly from the seed company, and most recently by the rapidly expanding use of TG seeds and the protections of IPR that they have been granted. In some countries (e.g., the United States) the latter lead to contracts under which farmers, when they have grown crops from TG seeds bought from an agribusiness firm, are legally prohibited from separating out seeds from their crops for subsequent plantings.[7] The commoditization of the seed, which depends on breaking the unity of seed (on the one hand) as source of a crop and (on the other hand) as reproducer of itself (Shiva 1997a), is an integral part of the transformation of the social relations of farming in the direction of the growing dominance of agribusiness and large-scale farming with, in many "developing" countries, export orientation. It serves corporate interests. Its proponents also maintain that it enables greater efficiency in agriculture (P_2) and, above all, that the farming methods associated with it enable much greater and cheaper production of the grains needed to feed the world's growing population (P_4). It serves, they maintain, not only corporate interests, but also interests pertaining to all value-outlooks.

Investigating Seeds under Materialist Strategies

P_1 encapsulates widely held views about the nature of scientific inquiry, including that research and development of TGs has been informed in an exemplary way by scientific knowledge or, rather, informed by exemplary scientific knowledge, knowledge gained under instances (biotechnological) of materialist strategies. If this is exemplary scientific knowledge, then investigating seeds qua biological objects has little to do with seeds qua constituent of a sociocultural location or agroecosystems, so that biology is sharply separated from studies concerning the sociocultural location. Whether particular seeds are commodities, renewable resources or gifts, objects with multiple roles in ecosystems, sources of marketable products or foodstuffs for local consumption, grown for the sake of multiple products or a single one, seeds (and plants)—qua objects of biological investigation—are effectively reducible to their genomes and to the biochemical expressions of their component genes. Then, the possibilities of seeds are encapsulated in terms of their being able to be generated from their underlying molecular structures (and the possibilities for their modification, which are expanded with the development of the techniques of genetic engineering) and lawful biochemical processes. Seeds are essentially as they are qua investigated in molecular, genetic, physiological, and cellular biology; that is, under biotechnological strategies (Lewontin 1992).

Understanding seeds biologically in this way thus (in conformity with biotechnological strategies being instances of materialist ones) largely dissociates the realization of their possibilities from their relations with social arrangements, with human lives and experience, with the social and material conditions of the research, and with extensive and long-term ecological impact (and with any other beings that might be recognized in a culture's world view)—thus, from any link with value. It deals with the decontextualized possibilities (section 1.3) of seeds. In turn, biological knowledge (so understood) is considered available to inform, more or less evenhandedly, agricultural practices regardless of the sociocultural location in which they may be inserted (P_2). Whatever seeds may become in agroecosystems (sustainable or high-intensive) is determined by the possibilities that are encapsulated in their genomes and the possibilities of their transformation (whether by natural or farmer-directed selection, or by bioengineering), and what they actually do become is brought about by means of chemical interactions with substances encountered in their immediate environments. There can be no feedback into (basic) biological investigation from considerations pertaining to seeds qua constituents of sociocultural locations. Rather, biological investigation comes first; only after it has been completed may the specific concrete uses of seeds, the interests they serve, and any other matters connected with value, be considered. Where the model "biology comes first; sociocultural analyses second" is assumed, the idea that the biological knowledge will be applicable evenhandedly across sociocultural locations will be well entrenched.

It is indeed true that sound scientific knowledge, accepted in accordance with impartiality, has been obtained under biotechnological strategies (attested to by the efficacy of TG technology) and that the identification of decontextualized possibilities of seeds (and of other constituents of agroecosystems) may be able to inform farming practices (in principle) in all sociocultural locations.[8] But, rehearsing my earlier arguments about neutrality (section 1.3), this does not imply that applications can be made evenhandedly or that actualizing socially a particular decontextualized possibility is valued across locations or that in all locations the relevant knowledge for understanding seeds is limited to (or even is predominantly) knowledge gained under materialist strategies. The knowledge embodied in FS seeds is (for the most part) not of this kind. The sociocultural location may provide feedback about the relevant kind of strategies to adopt in research.

Investigating Seeds under Agroecological and Related Strategies

Sound knowledge, relevant to farming practices, can be obtained under strategies that are not reducible to materialist strategies. A lot of sound knowledge is embodied in FS seeds (Altieri 1995); they are, after all, the sine qua non of the development of TGs. This illustrates the fruitfulness of research conducted under agroecological strategies (section 5.4), which (I have pointed out) is important for those who hold the values of popular participation (section 6.3). Other arguments have also been made claiming that it is shortsighted to consider research conducted under materialist strategies to be uniquely exemplary. They will be discussed in section 10.1, where it is shown that there is considerable evidence that research conducted under agroecological (and closely related) strategies can be fruitful. (This is important for evaluating P_1. There would be good grounds for endorsing P_1 if it were shown that fruitful research were not generated—after a serious effort—under competing strategies that are proposed by various groups.)

Shiva discusses this issue extensively (e.g., Shiva 1991). First, (drawing on many of the same sources that I cite in this book and others) she reiterates the productivity, the potential for increased productivity, the agroecological soundness of many traditional agricultural practices, and the possibilities for enhancing agroecological systems in the light of systematic research conducted on them. Secondly, she questions the efficiency of the green revolution (compared with potential developments of traditional methods) in view both of the extensive and expensive chemical and other inputs needed to produce the higher yields, and of (she alleges) exaggerated claims about productivity gains since the actual gains made concern only a single crop and have been achieved at the expense of reductions in other products of traditional farms. Thirdly, she lists an array of social shortcomings of the green revolution (cf. Tilman 1998; section 10.1), including displacement of traditional small-scale farming, causing social dislocation (and consequent violence) and hunger among the communities that sustained it, loss of the knowledge that informs that kind of farming, and deepened dependence of conditions and possibilities in the "developing" world on the interests of the global-market (see also *SVF*: chapter 8; Rosset 1999; 2001).[9]

Shiva (and the other critics cited in this chapter and in chapter 10) recognize, of course, that, using the methods of the green revolution, the world's food production has increased dramatically over the past four or five decades, so much so that now enough food is produced to feed everyone in the world. The criticisms of the green revolution would not amount to much, unless the critics are right that there are other methods that, given appropriate resources for their research and development, would generate comparable or greater productivity. The green revolution actually increased food production; Shiva questions its necessity and denounces its wider effects (see also Shiva 1997b; *SVF*: ch. 8). One would expect proponents of P_1 to respond: Given that the green revolution was a great success, so what there may have been other ways? The important thing was to get the job done! To them, Shiva's claim—that productivity gains, comparable to those achieved in the green revolution, could have been made with appropriately developed traditional methods—appears to be empirically unwarranted, idle second-guessing. But, Shiva's counterfactual claim does not stand by itself. It is essential to her argument, but for her it is not just that the productivity gains could have been achieved by other methods; rather, after investigating the social and ecological effects of the green revolution, she casts doubt on its value. In doing so, she points to sources of alternatives to the biotechnological projects that are being hailed as successors to the green revolution (section 10.3) Her claim may be mistaken, but it certainly is not idle.

Shiva's (and the other critics') arguments challenge endorsing P_1. They suggest—translating their proposals into my terminology—that, in order to investigate the productive and other possibilities of alternative approaches to agriculture (e.g., agroecology), strategies that are not reducible to materialist strategies need to be adopted. Moreover, they suggest that, unless strategies are adopted that take into account that seeds, plants, crops, and their products are socioeconomic, as well as biological, objects (most TGs are commodities), research will not be able to take into account the full range of the side-effects (risks) of using them. Thus, strategies not reducible to materialist ones (and including agroecological ones) need to be adopted in addition to materialist strategies, in order to investigate P_4/C_4 (alternatives) and P_3/C_3 (risks) adequately.

Furthermore, research conducted under a variety of strategies is pertinent, and necessary, for addressing P_2/C_2 (benefits). One of the alleged benefits of using TGs is that it can contribute to alleviating hunger. Alleviating hunger was also the principal goal of the green revolution. Yet, even with it, massive hunger persists throughout the world. Simply producing adequate quantities of food is not sufficient to ensure that all of the poor are fed. Consider:

> In order to put an end to hunger, it is necessary above all to improve the distribution of income. Without furthering economic equality, without effective means to combat poverty, increasing agricultural production can only lead to an increase in the export of foodstuffs, something that is already occurring on a large scale in underdeveloped countries—a fact that is prudently omitted in the arguments of the defenders of transgenics. (Alves da Silva 2000, my translation)

In order to eliminate hunger, it is necessary to eliminate its fundamental causes (cf. section 8.3). This is not to say that techno-scientific developments are not relevant, but only that their relevance cannot be appraised independently of systematic empirical investigation of the causal network that maintains hunger, investigation conducted under strategies that can assess socioeconomic causal factors.

Shiva on "Reductionist Science"

Shiva refers to science conducted exclusively under (what I call) materialist strategies as "reductionist science" (Shiva 1988; 1997a). It is characteristic of reductionist science to offer understanding of phenomena exclusively in terms of their underlying structures and their (molecular) components, their processes and interactions, and the laws that govern them—in dissociation from their relations with human life and experience, and with social and ecological relations, so that the objects of reductionist science are per se devoid of any value, "dead, inert, valueless." It thus effectively considers things as reducible to their underlying structures (with, today, an emphasis on the genetic structures of organisms), and attends, for example, to the possibilities of seeds that may be realized by manipulations of their molecular components and their interactions with other objects understood in the terms of reductionist science (e.g., herbicides), without attending simultaneously to the effects on human health and the environment that may be caused by introducing such modified seeds into farming practices, and the social effects that may follow from the socioeconomic context of such introduction.

Reductionist science also tends to treat phenomena in a fragmentary way, as a set of traits that can be investigated individually—a crop, for example, is treated as a source of one product (grain) and investigated as such, leaving aside that crops may also be sources of fodder for cattle or thatch for roofs of houses, expressions of cultural values, means for nurturing biodiversity, etc.; vitamin A deficiency is addressed as an item by itself when considering the problem of malnutrition separate from its causes and accompaniments (section 8.3). Finally, since it articulates understanding in dissociation from the social relations of phenomena, reducing the seed to its underlying genomic structure, reductionist science helps to disguise the fundamental transformation of the seed that is involved with its commoditization.

According to Shiva, reductionist science is intimately connected with the logic of the expansion of the market, an instrument of neoliberal forms of globalization; its pursuit has no other rationale and it has no significance and no applicability outside of the logic of the market. On application (but also per se for it cannot be applied significantly to inform projects that further values like those of popular participation), reductionist science generates what she calls a "fourfold violence."

First, violence against the supposed beneficiaries of knowledge (e.g., poor farmers and their families). The conditions in which they can continue to practice their forms of farming are undermined, so that they cease to be producers of their own food and become consumers who must buy food and who, often, following social dislocation are unable to gain access to sufficient nutritive food for themselves and their families.

Second, violence against the bearers of "non-reductionist" forms of knowledge (e.g., that obtained under agroecological strategies and that embodied in FS seeds). Granting "monopoly" to knowledge gained in reductionist science and, even more so, granting privileged protection of IPR to products informed by it, devalues the knowledge of the bearers of other, traditional and agroecological, forms of understanding and the activities informed by them. It also poses no barriers to social and economic projects that freely exploit these forms of knowledge (biopiracy) or that diminish their practical relevance and thus the agency of their bearers.

Third, and directly connected with the second, "the plunder of knowledge," or violence against "knowledge itself" when non-reductionist knowledge is held not to be knowledge at all—and, in the name of sound "scientific knowledge," traditional knowledge is not only devalued but also exploited, suppressed, distorted, and not considered to be an appropriate object for further empirical investigation and improvement.

Fourth, "the plunder of nature," or violence against the "object of knowledge." Projects informed by reductionist science tend to "destroy the innate integrity of nature and therefore rob it of its regenerative capacity," or to destroy the biodiversity and genetic heritage and the social and cultural traditions of a region. Consider:

> Sustainable agriculture is based on the recycling of soil nutrients. This involves returning to the soil part of the nutrients that come from it and support plant growth. The maintenance of the nutrient cycle, and through it the fertility of the soil, is based on the inviolable law of return that recognizes the earth as the source of fertility. The Green Revolution paradigm [and also models that deploy TGs] substituted the regenerative nutrient cycle with linear flows of purchased inputs of chemical fertilizers from factories and marketed outputs of agricultural products. Fertility was no longer the property of soil, but of chemicals. The Green Revolution was essentially based on miracle seeds that needed chemical fertilizers and did not produce plant outputs for returning to the soil. . . . The activity lay in the miracle seeds, which transcended nature's fertility cycles. (Shiva 1997a: chapter 3)

Shiva maintains that it is of the nature of "reductionist knowledge" that it leads to the forms of violence she diagnoses—not just the particular uses to which this knowledge is put. She denies that, even under (what she would consider) desirable socioeconomic conditions, it could be used to further projects that do not occasion such violence. Reductionist knowledge necessarily serves the interests of capital-intensive agriculture. While I think that Shiva's analysis of the fourfold violence is acute and helps to explain why, from certain socioeconomic locations, there is interest in identifying strategies of research that do not reduce to materialist strategies, in my opinion it is a mistake to relate reductionist science directly to capitalist logic (Lacey 1999: Lacey & Barbosa de Oliveira 2001). That relationship, I think, is mediated by commitment to the modern valuation of control. The violence that Shiva diagnoses may be tolerated in view of commitment to the modern valuation of control—"it is the price of progress!" But some of the proponents of TGs (e.g., those linked with CGIAR—chapter 6: appendix) claim genuine "humanitarian" motivation; they see their projects, conceived of as essential in view of their commitment to the modern valuation of control, to be set apart from capitalist logic, while at the same time they often have a coincidence of interests with agribusiness (section 8.3). (I think it important to make distinctions among the proponents of TGs.) The modern valuation of control is separable from capitalist logic although at the present time, since neoliberal institutions are its foremost bearer, the two are not separated. (Recognizing this may provide an opening for constructive dialogue between the "humanitarian" supporters of TGs and agroecologists—section 10.4.) Shiva maintains that the motivation of both the green revolution, and now of using TGs, is to bring about decisively the capitalist transformation of agriculture and thus the commoditization of the seed (cf. Lewontin 1998; Lewontin & Berlan 1990; Berlan 2001). Although she may not be right about the motivation in general, she may be accurately describing the effect. When the investigation of decontextualized possibilities is separated from that of other kinds of possibilities, things are investigated in dissociation from the conditions for the realization of their possibilities; so it will not be part of the "technical" investigation to figure out the social conditions under which the possibilities may be realized and what their consequences may be—so it may be missed that to interact with a thing so as to realize certain of its decontextualized possibilities may actually be also to treat it as a certain type of social object.

Alternatives to Reductionist Science

Shiva holds that reductionist science provides one kind of scientific knowledge.[10] Agroecological investigation provides another, one that is the basis of sound agricultural practices of special interest to small farmers in "developing" countries, an essential part of their struggle to maintain and develop their cultural heritage as well as to meet their material needs. It also challenges the fourfold violence engendered by the methods informed by reductionist science and offers the hope of reversing this violence.[11] Agroecological investigation—while today it draws in countless ways upon knowledge of the underlying structures, chemistry and biochemistry of plants, soils, and inputs into agricultural production—integrally locates farming phenomena, and therefore the seed, within their specific ecological and sociocultural location, and poses questions that do not involve abstractions from it. It investigates relations and interactions between an organism and its environment in a way that enables us to identify the possibilities that things (e.g., seeds) have in virtue of their place in agroecological systems. Unlike reductionist science, agroecology does not dissociate from the social, human, and ecological dimensions of things. Its focus is upon objects—productive and sustainable agroecosystems and their constituents (seeds, plants, microorganisms, etc.)—whose possibilities cannot be reduced to those identified with reductionist methods. Producing a crop is seen as a part of generating and sustaining productive agroecosystems.

Shiva emphasizes that in agroecology important understanding can often be gained from improvements (based in empirical research) of traditional farming practices and understanding, which themselves may be understood as part of agroecology, forms of investigation in which phenomena are not abstracted from the ecological and social relations that they exhibit. She points to the empirical success of agroecology (cf. section 10.1) as an important ground on which to challenge the monopoly on knowledge made in the name of "reductionist science." She points to the empirically vindicated strengths of local, traditional, people's knowledge, which, because of its locality, assumes numerous, diverse forms. Informed by such knowledge, traditional farming has, in some cases, developed practices that are ecologically sound (maintaining, e.g., soil that has remained fertile for millennia and pest and disease controls that function through appropriate arrangements and combinations of crops), selection processes that have generated a richly diverse gene stock, and modes of social organization in harmony with natural processes. In some writings, she has referred to the relevant traditional knowledge as dealing with "preserving and building on nature's processes and nature's patterns," with "repairing nature's cycles and working in partnership with nature's processes," and with "subtle balances within the plant and invisible relationships of the plant to its environment" (Shiva 1991: 26, 29, 97) Moreover, she points out, it is open to improvements (including those that derive from the input of knowledge of the underlying structure, processes, interactions, and laws of agroecological systems) with research, in which local farmers as well as "specialist researchers" would be active participants. Clearly, over the centuries, indigenous seeds (obtained from free pollination in open fields) have been improved as a result of the selection practices of local farmers. Their knowledge can become an object of systematic investigation in which it is articulated, systematized, empirically tested, and further improved.

NOTES

1. For example, "It [genome sequencing—including of the plant *Arabidopsis thaliana*, that "has quickly become the laboratory mouse of the plant world, studied for insights that can be

applied to virtually all other plants (Pollack 2000) and the bacterium *Xylella fastidiosa*, that cause diseases in many plants of agricultural importance] might well be the breakthrough of the decade, perhaps even the century, for its potential to alter our view of the world we live in. The pace has been frantic. . . . [The] lure of this knowledge has made the quest irresistible" (Pennisi 2000). "Where were you on 26 June 2000, while history was being made? On that day, at press conferences throughout the world, scientific leaders ceremonially opened the 'book of life,' announcing the completion of a working draft of the human genome" (Aldhous 2000).

2. Using words from an actual religious tradition, Prince Charles countered with "seeds of disaster" (Prince of Wales 1998).

3. Compare the discussion of the U.S. trade lobby in chapter 6: appendix.

4. The extreme case of biopiracy occurs when a foreign agency gains patents on minor variants of products that have been available for centuries in "underdeveloped" countries and that are well understood within local knowledge systems—for example, products of the *neem* tree in India (Shiva 1997a: 69–72) and basmati rice (Shiva 2000a: 84–86). (On the role of biopiracy in certain areas of medical research and the pharmaceutical business, see Alier 2000.)

5. The notion of "full understanding"—and its contrast with "wide-ranging" understanding—is introduced in *SVF*: chapter 5.

6. In this section I draw extensively on the works of Shiva.

7. Other countries (e.g., Argentina and Brazil) are under pressure from the WTO to introduce legislation that will require conformity to this practice.

8. I pointed out in section 5.4 that agroecology makes use of knowledge gained under materialist strategies, for example, in identifying bacterial and mineral components of agroecosystems; and genomic analysis may be a useful aid to traditional practices of selecting seeds (Miguel Guerra: personal communication, March 2, 2004).

9. Criticisms of high-intensity models of agriculture, similar to those made by Shiva, are common among movements of small farmers in many "developing" countries. I have noted some in recent documents of the Brazilian movement, MST (chapter 6: appendix). Here are two that I draw from my file of MST documents: Referring to "a North American type model that was adopted in Brazil," "In the past ten years the number of small-scale farmers has been reduced (by 95,000) and of rural workers (by 200,000). Accompanying this, Brazil has become an importer of foods; and, as incredible as it may seem, today it even imports fresh corn" (Gilmar Mauro, an MST leader, in a document, "Globalization, ecology and the small farmer"). And: "Illiteracy, violence, deforestation, contamination of foods and waters, destruction of soils, intoxication of workers, and a whole generation of landless persons are the marks that an agrochemical model has left and continues leaving in our country" (in a document, "MST and the environment: Agrarian Reform is a way to take care of the environment"). Both of these quotations (my translations) are referenced in Lacey & Barbosa de Oliveira (2001), where additional references can be found.

10. In this subsection some sentences from section 5.4 and section 5.5 are repeated, or paraphrased into Shiva's terminology, so that the affinities of Shiva's and Altieri's ideas will be apparent.

11. Once again, we find interesting parallels in the thinking of MST. Consider: "Illiteracy, violence, deforestation, contamination of foods and waters, destruction of soils, intoxication of workers, and a whole generation of landless persons are the marks that an agrochemical model has left and continues leaving in our country. In various settlements of the Agrarian Reform organized by MST, however, this picture is changing. Following the guidelines of

agroecology, seeds, agricultural inputs, and food are being produced in a new way. We can point to the intense effort of those in the settlements to produce seeds from their crops, retrieving varieties that had been previously utilized and that are well adapted to particular regions, but which were discarded by the firms that produced hybrid seeds. They are even producing (without using poisons and chemical fertilizers) seeds that are serving as the base for the totally organic and natural production of greens, vegetables, and fruit. These seeds are marketed by one of the cooperatives of MST under the brand *Bionatur*" (quote from Lacey and Barbosa de Oliveira 2001, my translation).

Chapter Eight

Benefits of Using Transgenics

P_2 There are great benefits to be had from using TGs now, and these benefits will greatly expand with future developments, among which are promised TG crops with enhanced nutritional qualities that can readily be grown in poor developing countries so that TGs may become key to addressing problems like those of hunger and malnutrition. When these promises are fulfilled, the benefits of TGs will become spread evenhandedly so as (in principle) to serve the interests and to improve the farming practices of groups holding any viable value-outlooks.

C_2 The benefits claimed for currently used TGs reflect the ethical/social values of agribusiness, large-scale farmers, and others who are beneficiaries of the global market. Furthermore, not only are the benefits relatively slight (perhaps even exaggerated by the proponents), being confined largely to these groups and not extending to small-scale farmers in the "developing" world (or to organic farmers in the advanced industrial societies), but also the promises made about future benefits are not credible, in part because developments of TGs reflect the interests of the global-market system, the very same system within which poverty, the fundamental cause of hunger and malnutrition, persists today.

Issues about the benefits to be gained from TG-oriented agriculture, and for whom they can be considered benefits, have already been discussed to some extent in the previous two chapters. The developers, producers, and those who choose to use TGs maintain that there are significant benefits, which clearly they value, to be gained from growing the currently available TG crops. That fact leaves unanswered, however, the following questions: How valuable are these benefits? Can we expect greater benefits from further innovations of TGs? For whom, and with respect to what value-outlooks, are uses of TGs considered valuable? Is the value for the short or long term? Is it sufficiently likely that the benefits will actually be gained in sufficient quantity and in a way that could not be matched or surpassed by alternative agricultural approaches? In the long run, of course, these questions are intertwined with others about risks and alternative possibilities (next two chapters). Furthermore, any discussion about benefits is necessarily implicated in (ethical and social) value judgments; what is an object of value within the context of one value-outlook may be negatively valued in another. Then, in view of the thesis that there are mutually reinforcing relationships between the value-outlook one holds and the research strategies one adopts, it might appear to be unlikely that TGs will be considered objects of value (to have benefits) apart from the value-outlooks to which are linked the strategies that shaped the research that produced TGs.

Nevertheless, in responding to the questions posed in the previous paragraph, the P-side (last sentence of P_2) affirms the neutrality of the knowledge that informs developments of TGs. It portrays TG-oriented agriculture as having value that transcends the current sociocultural location of its use and as following a trajectory that promises benefits for everyone, regardless of their value-outlooks, including poor farmers and their communities. In this chapter, in which P_2 and C_2 are the reference points, I will look at the arguments made for and against the claim for the universal value of TG technology and the neutrality of the scientific knowledge that informs it.

8.1 CURRENT BENEFITS

One might wonder about the neutrality of the scientific knowledge that is embodied in TGs. At present most TGs are also commodities (section 7.2). Many factors contribute to ensure this. Most research on TGs has been, and continues to be, conducted in corporate research institutions and university departments whose research is funded by or linked to the interests of agribusiness. Agribusiness also is responsible for most of the actual implementations of TGs, and it has legal control—furthered by the granting of IPR (e.g., patents)—over not only many TGs themselves, but also over the techniques and procedures of genetic engineering and even over certain genes and plant characteristics. Moreover, the spread of TG-oriented agriculture to "developing" countries, covered by the protections of IPR that are backed by the WTO, is an integral component of current neoliberal programs of "globalization." In this context, it seems, any products of research and development (R & D) conducted under biotechnological strategies, exploring the possibilities encapsulated in the genomes of seeds and the possibilities for modifying them, will almost inevitably become commodities, so that there seems to be no place for the use of TGs outside the sociocultural locations shaped by market relations. Thus to engage in this kind of R & D is simultaneously to contribute to the interests of agribusiness and the market, and it has little relevance to the projects of those who are attempting to improve productive and sustainable agroecosystems that use methods that are in continuity with traditional local knowledge (section 7.2).

This is not surprising. In the development of TGs, the model "biology first, sociocultural analyses second" (section 7.2) was not followed. TGs were introduced into agricultural practices, not because there was a consensus of scientific opinion that it was vital to introduce them, but because of the interests of agribusiness (and those of its clients, neoliberal institutions, and their state backers). As indicated above, R & D of TGs has largely been under the control of agribusiness, and increasingly the very objects of research are themselves patented objects—and thus owned objects; and patents have no meaning outside of relations shaped by property and the market. This is not to deny that research on TGs has led to the discovery of genuine scientific knowledge, that genuine and realizable possibilities for the genetic engineering of seeds, which enable them to produce crops with certain "desired" characteristics, have been identified. If this were not so, there would be no efficacious applications and thus no controversy. TG technology is informed by soundly accepted scientific knowledge—knowledge that is grounded solely on the empirical evidence and the play of the cognitive values, and that accords with impartiality. But, we have seen, impartiality does not imply neutrality (section 1.3). Keep in mind, however, that there are cases where a line of investigation was pursued because it was expected to produce knowledge of interest for a particular value-outlook, but where this knowledge also came to serve the interests of other value-outlooks (see below, section 8.3)

Agribusiness corporations, at the present time, are the primary agents and beneficiaries of the development and deployment of TG seeds. The principal objectives of agribusiness are profit and, relatedly, gaining greater control of the market and guaranteed larger sales of associated products. Since producing profit is not among the decontextualized possibilities of TG seeds, it is not investigated in the biological or technical research. The question, "Which of the decontextualized possibilities would, on realization, generate profits?" is not a question for biological research. Nevertheless, research on TGs has been framed and oriented by answers to it (illustrating my remarks on the unity of the natural and social sciences in section 2.1). Possibilities for profit require that certain social relations and economic factors be investigated. This can be quite a fragile matter and subject to rapid changes with varying vicissitudes of the times. The future of TGs may well be influenced, not so much by considerations of their efficacy, legitimacy, or humanitarian value, but by the market, for example, that Europeans will not buy TG products. There can be no profit if there is no market for the products, and presumably there will be a market only if potential buyers see benefits in the use of the products.

Markets are created, in part, through advertising, and it is part of the "discourse" of advertising to present products in the most favorable light possible, thus ignoring the nuances and qualifications that should accompany any "scientific" discussion of alleged benefits. We cannot take agribusiness claims about benefits at face value, since they mainly come from the declarations of corporate spokespersons and the rhetoric of those who are seeking funding for their research projects. I caution that agribusiness may easily blur the line between science and advertising.[1] Nevertheless, there are genuine benefits anticipated for some of the users of TGs, including industries that process and market agricultural products and the farmers who have available the conditions to use them. Certainly the typical kinds of TGs currently used—RoundUp Ready and similar plants, and plants containing Bt toxin—do (or are expected to) contribute profits to agribusiness and its clients. Moreover, presumably corporations aim to produce particular kinds of TGs that will afford them some competitive advantage. It has been asked: Why did Monsanto Corporation develop RoundUp Ready seeds? Was it because the active RoundUp ingredient (glyphosate) is the most effective and least environmentally intrusive of known chemical herbicides, or because Monsanto had controlled patent rights to glyphosate and wanted a means to ensure a continued protected market for the highly profitable pesticide? Whatever Monsanto's motivation may have been, claims about other kinds of benefits should not be ignored, but submitted to proper investigation.[2]

Using available TGs may contribute to ease certain burdens of and worries about large-scale, conventional high-intensity farming. In the words of Paul Thompson: "The tools of science that we know as food biotechnology can be employed to increase agricultural productivity, reduce negative environmental impacts and to ensure and improve food safety" (Thompson 1997: 18). For the farmers who use them, it is said that there are such benefits as less, and less frequent, use of herbicides and chemical pesticides (and thus less personal exposure to toxics and lessened likelihood of health problems caused by such exposure), and thus lower costs of inputs, easier work, fewer lost crops, higher yields, and higher income. In turn, this is said to underlie benefits for the environment: less pollution from herbicides and chemical pesticides and smaller areas of growing space needed (thus putting less stress on the environment); the growing of TG crops is "environmentally friendly," and may contribute to reversing some of the environmental damage caused by chemical-intensive (including green revolution inspired) farming. Many farmers tend to prefer this kind of farming to conventional farming, in significant part because it is easier and safer. (This probably helps to explain why many farmers in southern Brazil have been willing, despite the risk of prosecution, to use TGs illegally smuggled from Argentina.) As already indicated, conventional farming is chemical-intensive and typically requires repeated and intensive applications of herbicides and pesticides, which can cause great harm to health. The problems (threats to the health of farmers and environmental devastation) of conventional farming are real—and the claim that TG seeds represent a way to avoid these problems is speaking to a genuine and urgent concern; and it is an important factor for farmers (who operate reasonably large-sized farms), who have readily adopted the new seeds (Pollan 1998; Rauch 2003; Thompson 1997). There are also said to be benefits for consumers: lower prices, foods with better tastes, greater variety of foods at all times of the year, even safer food—though it is generally conceded that these consumer benefits have yet to be realized on a significant scale;[3] and the economy in general is said to benefit from resulting enhanced economic competitiveness (Nuffield Council on Bioethics 1999: 5.22).

Nothing in the above listing of benefits undercuts the argument of the previous chapter that the knowledge that informs TG technology is not significantly applicable to serve interests shaped by the values of popular participation, and that today it has little relevance to the projects of those farmers who are attempting to improve productive, sustainable, and biodiverse agroecosystems that use, for example, agroecological methods. Although it serves powerful and widely shared interests, the value of gaining this knowledge is not universally upheld. This is not to deny, I repeat, that TG technology is informed by knowledge that is grounded properly on reliable empirical evidence. The cognitive value of this knowledge remains intact; its social and ethical value is questioned from the perspective of some value-outlooks. The C-side will recognize in the claims about environmental benefits a response to criticisms of the green revolution, like those made by Shiva (section 7.2). It will also remember that the chemical-intensive methods of agriculture that today are conventional were fostered by the same corporations that are now proposing methods to overcome the problems caused by conventional agriculture. Intensive use of chemicals once served the profit interest well; now the corporations want to move to the next stage that science has opened up for us. The environmental consequences of chemical-intensive farming were not foreseen—or at least not announced loudly in advance—by the scientists who conducted the research that informed these methods (including those of the green revolution). This invites the response from the C-side: "The science, applied in corporate projects, has been 'suspect' in the recent past; why should we take it at face value today?" The sting is removed from this response wherever the modern valuation of control is held—recall its presupposition (b): "Techno-scientific solutions can be found for virtually all problems, including those occasioned by the 'side effects' of techno-scientific implementations themselves" (section 1.1).

8.2 ANTICIPATED BENEFITS

Without benefits from current uses of TG seeds like those listed above no farmers would be planting them today. Even so, these benefits are at best quite modest and not widely shared among the world's population; certainly they would not be sufficient to sustain the legitimacy of these uses if seriously harmful side effects of sufficient magnitude were to be demonstrated, or if other approaches to agriculture showed greater promise for being able to address current and future problems of food scarcity. This reinforces the fact that arguments discussed in this chapter remain incomplete until supplemented by the discussions about risks (chapter 9) and alternatives (chapter 10). The argument that the benefits of using TGs will (can) be spread evenhandedly, however, goes beyond consideration of current benefits to that of anticipated ones.

The P-side counts a lot on redeeming its promises about the benefits of TGs (currently actively under development or in the planning stage) that involve more sophisticated methods of genetic engineering (see section 9.3). Techniques for producing TG seeds are relatively new. Currently available techniques are "first generation," and developments are taking place at a rapid pace. So, current methods cannot be taken as indicators of what the innovations of the future will be like; the possibilities of tomorrow are expected to surpass by far those of today—and so too, according to the P-side will their benefits. Hence, the P-side concludes, the value of the future products of R & D on TGs cannot be extrapolated from the value of currently available TGs. Current uses of TGs, they may indeed concede, are principally for the benefit of agribusiness and relatively large-scale farmers, since they generate crops that are resistant to pesticides produced by a corporation or that are grown from seeds over which the corporation has gained a virtual monopoly. But that reflects only that current methods are very simple. When the more sophisticated methods are introduced, the P-side says, more significant and widespread benefits will be forthcoming; and these will be a consequence of one current benefit that I have not yet referred to: viz. profits from current uses help to finance further R & D of TG seeds; and, furthermore, some of the benefits promised from these developments are apparently of great importance to all people, including small-scale farmers and the poorest of the poor. Look at the list of projected benefits! It includes TG crops that will display the following types of "desired traits":[4] higher yields; more nutritious grains, for example, vitamin-enhanced rice ("golden rice"); tolerance for salt, heat, frost, drought, altitude, mineral deficiencies, and other hostile environments that normally inhibit crop production; resistance to bacterial and viral disease, pests, and chemicals (herbicides, pesticides, fertilizers); grains with components that contribute to health, for example, lowering cholesterol; capability (in crop plants) to fix nitrogen in soils; capability for nitrogen-fixing plants to produce equivalents of certain nonnutritious agricultural products (e.g., coffee); "enhancements": aesthetic appeal of foods (flavor, texture, appearance), or market role (longer shelf life, more resistant to the stresses of travel, easier to harvest mechanically).

The rhetoric can be very persuasive. Here are apparently limitless possibilities (illustrated with some apparently compelling examples, e.g., "golden rice"), which challenge our imaginations, put forward so as to overcome all doubts about ethical legitimacy. (Who could not regard some of these traits as desirable?) It is as if the P-side were saying, "Tell me what you want produced to allay your doubts, and I'll assure you that we are on the way to realizing it. We are at the beginning of a process that will take us well beyond the limits of what has previously been considered possible or even imaginable. We are, for example, working on corn that will fix nitrogen in the soil. If this lies within our compass, how can you suggest that there are serious limits to what we can produce?" Also, reflecting commitment to the modern valuation of control and its presuppositions, it seems to be saying, "Don't be worried about risks—future technical developments will take care of them!" This rhetoric puts all the attention on technical biotechnological matters, and turns attention away from the socioeconomic mechanisms that are at the heart of the project. Yet, it is these mechanisms, and not just or even principally complex technical matters, which provide orientation for developments of TG and pose barriers to other approaches. When attention is focused on technical matters, we ignore the socioeconomic mechanisms that underlie who gets fed and who does not (and how this may be related to the methods used in production). To get at what is involved here I turn to a detailed discussion of "golden rice." It has captured the public imagination and the P-side often builds much of its argument around it; all the issues come to a head in connection with it.

8.3 GOLDEN RICE

"Golden rice" is rice that has been genetically engineered so that its endosperm contains beta-carotene, a source of vitamin A when ingested by human beings (Ye et al. 2000; Potrykus 2001; Guerinot 2000). ("Golden" refers literally to its color, a consequence of a daffodil gene being engineered into the rice genome, but play on its common metaphorical meaning, which frequently occurs in the press, seems to be meant to entice us to think that it is somehow outrageous to deny the value of "golden" rice.) Malnutrition is endemic in poor countries. Vitamin A deficiency affects millions of children in countries where rice is the staple food and is responsible for large numbers of deaths and cases of blindness. Golden rice is proposed as a contribution towards alleviating vitamin A deficiency, and thus saving lives and rescuing children from blindness in poor countries. To question the value of golden rice seems, to its developers, to be preposterous and heartless (unethical), not a matter of rational argument but only of "emotional appeal," to reflect only "anti-science" and "anti-market" sentiments (Borlaug 2000; Potrykus 2001).

Keep in mind, however, that the C-side is not refusing to accept a demonstrated "good" that is already at hand. Golden rice is not yet available for agricultural use; indeed, the first experimental plants were produced only a few years ago. That it will become available for the use of small-scale farmers is a promise that confronts many obstacles, whose redemption lies (if at all) some years off in the future. Meanwhile, as field tests and risk assessments are carried out, its inventors intend to develop golden rice, crossed with varieties currently grown by small-holding farmers, so that it can be grown by them in essentially the same way that they currently grow their crops. It will be given without charge, by the institutions that are developing it, to poor farmers who desire to use it, and it can be sown every year from seed saved from the harvest. Looked at this way, golden rice appears to be a regenerable resource and not fundamentally a commodity. Thus we have, in the words of its inventor, an instance of "the purely altruistic use of genetic engineering technology [that] has potentially solved an urgent and previously intractable health problem for the poor of the developing world" (Potrykus 2001). The proposed "humanitarian" uses of TGs are not, of course, limited to golden rice; research is under way to produce—among others—iron-enriched rice, crops that will grow in drought-afflicted regions, and crops resistant to chronic fungal or viral infections.[5]

Humanitarian As Well As Market Value of TGs

The developers of golden rice emphasize the humanitarian value of their projects. This seems to support at least a measure of neutrality of research on TG (humanitarian value, as well as value for agribusiness and its clients) and perhaps its universal value. Agribusiness spokespersons emphasize the neutrality and the universality (Leisinger 2000). Given the humanitarian value of some of its uses, they suggest, it follows that the development of TGs in general is desirable and so in particular is that carried out by agribusiness—the ethical value of the whole project of TGs is vindicated by the humanitarian value of golden rice (and a few similar cases).

The argument underlying this suggestion may go something like this: Agribusiness corporations, as a consequence of the R & D they have sponsored, hold the patents to certain genes and techniques of genetic modification that are needed to produce the seeds that serve humanitarian ends. So without the R & D that they have sponsored, no TG seeds can be produced. Thus, to deem their R & D undesirable implies deeming all developments of TGs undesirable. Agribusiness, of course, aims for profit; but at the same time it cooperates with humanitarian projects that have nothing to do with profit by licensing, without charge, use of their patented material for developments and implementations of seeds used in these projects.[6] So, although the immediate innovations of agribusiness are not designed to meet the food and nutrition needs of the world's poor, the R & D that underlies them is essential for, and part of, projects that are so designed. In this way the humanitarian value of innovations like golden rice is taken to support the value of TG technology in general. The developers of golden rice endorse this argument: for example, "One can only hope that this application of plant genetic engineering to ameliorate human misery without regard to short-term profit will restore this technology to political acceptability" (Guerinot 2000: 243); compare "Thanks to the interest of the agbiotech companies to use 'golden rice' for better acceptance of GMO technology, and thanks to the pressures against GMOs built up by the opposition, the IPR situation was easier to solve than expected" (Potrykus 2001: 1159). Minimally, there is a coincidence of interest between agribusiness and the "humanitarians" in the social and political acceptability of TG technology—a technology that both parties accept can be used in service to either of their interests.

Although the "humanitarians" sometimes question the propriety of particular emphases of agribusiness (e.g., Serageldin 1999), since their own developments are dependent on the products (and "charity") of those made by agribusiness, they do not question the policy of increasing the profile of TG-oriented agriculture that is furthered through the mechanisms associated with IPR. Thus the projects of the "humanitarians" do not challenge those of agribusiness or pose any obstacles to their spread. Rather their projects are intended for niches that agribusiness has no (current) interest in filling. They have a subordinate and dependent place in the process of market-oriented "globalization" of which agribusiness (and other major players of the worldwide market) set the direction—a place, moreover, that is contingent upon the need for agribusiness to seek public legitimation by appeal to the availability of its technologies for use in practices that serve humanitarian ends and by its own charitable contributions to these practices. The advent of golden rice, therefore, does not further the case that TGs have value outside of the socioeconomic location in which market relations are encompassing. This would not follow, of course, if there were no alternatives. (Once again P_4 is key to the argument.)

Skepticism about Golden Rice

This conclusion seems to clash with the conclusion of the developers of golden rice that it can be grown from year to year from seed saved from the harvest; that the one-time gift of the seed from the humanitarian research institutes would thus enable farming to continue productively without further dependence on outside institutions. The C-side is sceptical of this claim for several reasons (e.g., Ho 2000b). In the first place, the likelihood of any variety of golden rice being able to breed true for several generations is not high in view of the instability of engineered seeds (and the unsustainability of their crops), especially of those like golden rice where the engineering involved is complex and intricate. Second, unless golden rice were to become part of sustainable agroecosystems, periodically there would need to be sources of variation in the crops grown in order to cope with pest and other problems, and these sources would have to be the research institutes. R & D of golden rice requires an expert elite; it is not a project that wholeheartedly supports self-reliance or mutuality. Third, golden rice targets one dimension of malnutrition, vitamin A deficiency. If the way to deal with malnutrition is through producing (via genetic engineering) more nutritious crops, other genes will also need to be inserted into rice (and other crops) to deal, for example, with iron deficiency, so that one would expect a regular stream of new "improved" seeds from the research institutes (though the first source of skepticism would be exacerbated), replacing those that supposedly could be saved from the current crop and planted for the next crop. Fourth, sustainability is not built on the necessity of receiving regular "charity" from agribusiness, for (among other things) the interests of agribusiness may change. The seeds of golden rice, like virtually all TG seeds developed within the current socioeconomic context, remain entities to which IPR are claimed, so that the rights to their distribution, reproduction, and further modification, as well as to their straightforward commercialization, lie with those who hold the IPR. Even when those rights may have been licensed without charge to the farmers who use the seeds, the licenses are subject to being revoked or not extended. Although TGs might have been developed under different socioeconomic conditions, it does not follow that the technologies of genetic engineering currently developed would have significant applicability under different conditions, especially if those conditions subordinated TG technologies to sustainable agroecosystems and the enhancement of local communities and to their addressing problems that cannot be addressed by agroecological methods. (R & D conducted in a context where numerous IPR claims have to be negotiated may have little applicability outside of the realms where IPR regimes are recognized.)

I regard all these as matters for further investigation. The C-side sees a recipe for growing dependence on agribusiness or for creating conditions that will facilitate the extension of agribusiness's control over food production. If the commercialization of golden rice were to turn out to be profitable, then the companies might, for example, undersell the small-scale farmers, contribute to driving them out of business, and thus to opening up their lands for commercial exploitation.[7] These matters also pertain to the further criticism that may be made of P_2: that there are indirect effects on health and the environment that are mediated through the control of agriculture increasingly being gained by agribusiness, which can be expected to subordinate meeting the food needs of people, the interests of small farmers, and the nurturing of biodiversity, to its own objective of profits (see section 9.5).

In addition to skepticism about the P-side's claim that golden rice will become available for use, and use from year to year, by small-scale farmers, a variety of further objections have been raised questioning the value of developing golden rice. I consider some of them in the following subsections.

Golden Rice as a Techno-Scientific Solution to a Social Problem

Recall that golden rice has been proposed as a solution to a problem, vitamin A deficiency (and its accompanying devastating effects on health and well-being) among large numbers of children in impoverished countries. To its developers, that is the important thing. They are proposing a techno-scientific solution to it, and they consider issues connected with IPR to be peripheral. Their interest is gaining access to the means to put this solution into effect "realistically" under prevailing conditions (and that means constructive collaboration with those who hold the relevant IPRs). That prevailing conditions happen to be neoliberal is not the responsibility of scientists; but, if there is no other realistic way to put the solution into effect, then subordination to the neoliberal project is appropriate, even obligatory. Not to proceed with these collaborations is to fly from responsibility to solve the problems of the poor. Then, criticisms of golden rice, linked with issues of IPR and the key role that TGs are coming to assume in the global economy, are deemed irrelevant—or worse, as actually opting out of addressing such problems of the poor as malnutrition and hunger (Potrykus 2001; Borlaug 2000).

The language used by the proponents in hailing golden rice (and dismissing critics) reflects the modern valuation of control and its presuppositions. The C-side responds, however, that this techno-scientific solution has been proposed to the problem of vitamin A deficiency, without careful empirical inquiry having been engaged in on the causal history of the currency and extent of the deficiency, or the causes of its actual continuance (Rosset 2001). It is not that the P-side offers no analysis at all of the causal history of the problem; it tends to see it as a consequence of scientific, technological, and economic "backwardness," exacerbated by such things as wars, corrupt governments, and the bad advice of environmentalists in the advanced industrial countries. The criticism is that the proponents have done no careful empirical inquiry on the matter, so that their analyses are little more than repetitions of the presuppositions of the modern valuation of control. (This would not be an important criticism if these presuppositions themselves were soundly accepted in the light of empirical inquiry.) This is important, for a problem of this kind cannot be solved without the elimination of the factors that cause its maintenance. But, in golden rice research, the problem has been characterized in a way that makes it tractable to "scientific" solution (in this case one deriving solely from research conducted under biotechnological strategies). This is achieved by a priori identifying the problem with one of its aspects (see the third source of skepticism in previous subsection), and its cause with one of its necessary conditions, one that can be characterized with categories deployable within materialist strategies, so that the cause of vitamin A deficiency is represented as insufficient quantities of pro-vitamin A in the food ingested by poor children—suggesting that the social explanation that people are suffering from vitamin A deficiency is the absence of the relevant technology to produce the right kind of rice. "Science" can then hope to deliver a solution by engineering beta-carotene into the staple food.

But if poverty is a major causal factor and the problem is malnutrition, of which vitamin A deficiency is just one component, then there is no reason to expect that solving vitamin A deficiency in the "scientific" way will not leave the principal causal conditions of poverty and malnutrition untouched. It may even exacerbate them if, as suggested above, developments of golden rice are dependent on agribusiness and, as is affirmed in C_2, they are integral parts of the socioeconomic system that causes the kind of poverty (and its attendant sufferings) experienced actually by many people in impoverished countries.

The C-side does not need to query (although some of its adherents do) the sincerity of those who make the "realistic" argument in favor of the necessity of finding techno-scientific solutions to problems like malnutrition, or to suggest that, in their own minds, they are subordinating the needs of the poor to the interests of agribusiness. These scientists want to bring the latest techno-scientific advances to help address the needs of the poor. The important criticism is that acting "realistically" in the manner suggested may not enable a solution to the problem to be found and brought into effect because it ignores the causal nexus of the problem.

Often one finds in CGIAR and its associates (appendix: 1, variants, pro; see also section 10.3) a heavily moral-laden rhetoric of "helping." We, they say, are using the latest science for the benefit of the poor; then, they charge the C-side with irresponsibility, immorality, and recklessness. They tend to portray the C-side representing affluent people from the "developed world" who prioritize environmental preservation and restoration (and other "self-serving" causes such as the disruption of the normal functioning of the "global" economy) at the expense of all other issues, including meeting food and nourishment needs of the poor. The C-side also, the portrayal continues, focuses on risks without attending to potential benefits, consequently exaggerating the amount of risk involved—then, seeing little benefit from TGs for their own interests, it downplays "real" benefits that can be expected for the poor; and its stance is "anti-science," "luddite," challenging the value of techno-scientific innovations, because it is caught up in nostalgia for a romantic, imagined past. (For different emphases on these themes, see Borlaug 2000; Potrykus 2001; Human Development Report 2001. On the "luddite" theme, see Lewontin 2001.) The rhetoric claims the authority of scientific expertise and the epistemic privilege of the experts. (But does P_4, or P_3, represent well-established scientific knowledge?)

Those who deploy this rhetoric tend not to ask: "Why is there hunger and malnutrition?" at least not in a context that would be open to the evidence of the historical record (and allow at least the possibility that previous programs of "helping" causally contributed to the current situation, or that the cause of hunger is the socioeconomic system which both supports and is furthered by the scientific research in question). Instead, it is taken for granted a priori that science (research conducted under materialist strategies) contributes to solutions and does not cause problems. (If previous programs did cause problems, then that was not because of the science but because of the administration of the programs.) Once more, the C-side notes, the "developed world" is looking to help the "developing world," totally discounting the need for local solutions to hunger and malnutrition. This rhetoric is part of the discourse of Western "moral superiority"— scientific advances serve "morally sound" causes. Yet, adherents of the C-side often say in frustration: "After all this time and all the programs, look at the condition of the 'developing world'!" They doubt that this time will be different, that with the new science of biotechnology the moral superiority (and the "generosity") of the West will finally be vindicated.[8]

Sometimes it is said that the opponents of TGs want to deny the poor in "developing countries" the choice of whether to use this new technology or not (Human Development Report, 2001). The point should be extended: Why not have poor farmers participate in the choice of what kinds of research are conducted, so that in practice they could confront the choice of TGs or well-developed agroecological (or other) methods? The fact that there are alternative ways does not mean per se that the implementation of TGs should stop; it does mean that research (at least publicly funded research) should not be limited to the one option of TGs.

Is Golden Rice an Adequate Techno-Scientific Solution?

Even assuming the successful development of golden rice in the future (Lewontin 2001, among others, doubts this) there have been questions raised about whether its ingestion would solve the problem of vitamin A deficiency. Some say that the amount of beta-carotene that could be ingested with daily intake of rice is insufficient to produce the needed amount of vitamin A, and that the body's capability to produce vitamin A depends not only on the ingestion of beta-carotene, but also on the presence of other factors—certain fats (Ho 2000b) and a generally well-nourished body (Lewontin 2001). These critics maintain that the possibilities of golden rice have been seriously misrepresented.

Others object that standard risk assessment procedures followed in connection with TGs (Potrykus 2000 points out that they are being followed with golden rice) are rationalized by the assumption of "substantial equivalence" of TG and farmer/breeder selected seeds (see appendix: 2), which cannot be justified. The genome of golden rice, for example, contains not only the genes for production of beta-carotene, but also other "foreign" genetic materials that are needed for several purposes: to insert the desired gene into the genome, to signal when the gene is to express itself, to function as bacterial markers, and several others. The risks that may be occasioned by these other materials are not all investigated in standard risk assessment procedures, and how serious they may turn out to be remains largely conjectural (Ho 2000a; 2000b). The concerns about risks would be heightened if the criticism stated in the previous paragraph were to be sustained. In that case, one might conclude that the principal role of golden rice is that of providing legitimation for TG technology or of gaining support and funds for scientific research in molecular biology and related fields.

Alternatives

I mentioned above that the practical use of golden rice remains a promise, the redemption of which lies (if at all) perhaps five or six years off in the future. If it had already been developed and implemented, then perhaps the objections would be able to be met by demonstrating the advantages empirically and showing that they were indeed available for the free use of small-scale farmers, or the empirical record might confirm the critics' worries. Thus, unless one holds the modern valuation of control, there is no compelling current ground to support the universal value of the project of developing golden rice (let alone of golden rice itself). More than this, the C-side, which holds generally that the empirical record points more in the direction of C_4 than of P_4, maintains that there is a positive case that there are specific better alternatives than developing golden rice (chapter 10).

That there are (potentially) other ways is the key critical contention, other ways without the risks that may be occasioned by TGs, which utilize agricultural methods (e.g., agroecology) that are informed by sound scientific knowledge and that are of special relevance to small-scale farmers in the "developing world" (e.g., Altieri 2000a; 2000b; 2001; Ho 2000b; Nodari and Guerra 2001; Shiva 2000b). These other ways address problems like malnutrition as a complex phenomenon with many interacting components, instead of as a set of discrete issues with each one awaiting its individual solution—in a context that does not separate proposed solutions from detailed analysis of the causal history of the problem, and that looks for solutions that are ecologically sustainable and supportive of the well-being and enhanced agency of members of the local community. If there are serious alternatives, then it is not enough just to work towards the hoped for redemption of the promise of golden rice (and related crops). The important issue becomes: What are the options now, and which ones are the most promising? The critics contend that the other ways they favor (at least as candidates for further investigation) can deal with vitamin A deficiency as one source of malnutrition among many. These ways involve farming in diverse agroecosystems that include a variety of greens, vegetables, and oil-producing plants that are sources of many minerals needed for human nourishment, and encourage uses of crops that are not wasteful, for example, consuming the husks of rice that do contain pro-vitamin A. These alternatives are well known and the relevant plants are locally available. Their implementation could begin immediately (unlike uses of golden rice) with relatively little costly research and development compared to what is required to develop golden rice (Altieri 2000a; Ho 2000b; Rosset 2001; Shiva 2000b).

Agroecology does not involve a "blueprint" which can be proposed for implementation widely throughout the world. It consists rather of a cluster of locally variant developments. Its potential can be tested locale by locale, in locales where farmers desire to experiment with it. It occasions none of the risks of a rapid large-scale transformation of agriculture; it is not another "revolution" to surpass the green revolution. It does not permanently threaten the conditions needed for other forms of agriculture. What is to be lost, for example, in experimental implementations of agroecology in areas afflicted with chronic vitamin A deficiency during the years spent waiting for golden rice even to be available to be planted by farmers?[9] There is a compelling case that now agroecology can play an important role in meeting the basic food and nourishment needs for many poor rural communities. It remains open to further investigation how extensive the potential contribution of agroecology is compared with other forms of agriculture. The case for adopting agroecological strategies is that they should be adopted as one among a multiplicity and diversity of strategies, as strategies whose products are especially likely to address interests of poor farmers in "developing" countries. This confirms that arguments for the benefits of TGs being universal presuppose that there is a compelling case against C_4.

If there are alternatives (like those discussed in section 5.4; section 10.1), then the value of the use of high-intensity forms of farming, including TG-oriented ones, cannot be assured universally, especially among those groups who experience the sufferings induced in the light of the social shortcomings that may result from them (section 7.2). Even if TG-oriented agriculture may be the only form of high-intensity agriculture that is viable in the long term, there are few who expect it to produce significantly more high-yielding plants than those currently used in high-intensive farming (Weiner 1990) or to be more environmentally sustainable than well-designed "ecological" methods. Thus, whatever its other merits might be, they do not preclude that, from the perspective of some value-outlooks, "ecological" methods may have higher value. Shiva anticipates that implementations of TG-oriented agriculture will exacerbate the shortcomings that she diagnosed with the green revolution (section 7.2), and she emphasizes their (alleged) inability to provide solutions to the actual problems of small-scale farmers (Shiva 1997a; see also Altieri 2000a). If this is so, then, at least among those who bear the brunt of the shortcomings, TGs will not be highly valued, especially if they belong to movements aiming to develop alternate modes of farming that are highly productive, ecologically sustainable, and protective of biodiversity (Tilman 1998), and also compatible with social and cultural stability and diversity (Shiva) or with the values that Altieri calls "sustainability" (section 5.4) or those of popular participation (section 6.3). Furthermore, the C-side maintains, the implementation of TG-intensive agriculture contributes to undermine such movements and their projects—by means of reliance on IPR claims, furthering the process of commoditization, and engagement in biopiracy, which, in turn, contributes to undermine the continued maintenance of seeds as regenerative resources (section 7.2; section 9.5).

Shiva (1997a) has summed up the disagreements about benefits (and risks) in a striking way. The seed, she says, has become a fundamental symbol in contemporary struggles. As commodity, it symbolizes the capability and power of the market—making use of technical innovations and legal mechanisms—to penetrate into realms that hitherto had resisted such penetration (cf. Lewontin 1998; 2001). As regenerative resource, it symbolizes the possibilities of local enhancement, agency, initiative, and well-being, of everybody being well fed and nourished, of cultural and biological diversity, of ecological sustainability, of alternatives to the uniformity of neoliberal institutions; of genuine democracy. Shiva writes: "The seed has become the site and symbol of freedom in this age of manipulation and monopoly of its diversity. . . . The seed . . . is small. . . . In the seed, cultural diversity converges with biological diversity. Ecological issues combine with social justice, peace, and democracy" (Shiva 1997a: 126).[10]

Issues about benefits are clearly essential to discussion of the legitimacy of using TGs, but they are not decisive. It also needs to be shown that risks occasioned by their uses (and anticipated uses) do not outweigh the benefits. Moreover, to develop a compelling argument legitimating that TGs should be accorded the predominant role in the public agricultural policies that will shape future farming practices, it would need to be shown that there are no viable alternatives that could generate comparable or greater benefits. Risks and alternatives will be discussed in the next two chapters.

NOTES

1. It is easy for corporate spokespersons to control the direction of the discussion, blending as they do the discourses of science and advertising. They need only to assert that there are certain kinds of benefits—and then the burden tends to be put on the critics to refute the claims (although generally they are denied access to the findings of corporate research). The critics are thus continually on the defensive and they cannot "win" the argument. In this context, some of the critics have had recourse to unruly, disruptive demonstrations, for which the P-side has not hesitated to brand them "anti-science" and "anti-progress." But blurring of the line between science and advertising and holding confidential the data of many corporate sponsored research projects strike at the integrity (impartiality and autonomy) of science in a much more profound way.

2. This can be a complex affair. Referring to one of the examples just below, even if using RoundUp Ready seeds may mean that generally fewer applications of herbicide will be made on a given field, it does not follow from this that, overall, smaller amounts of herbicides will

be used than previously, for a greater amount may be used in each application (it won't harm the crop), and it may be used more extensively (or more frequently) over larger areas. Claims about benefits need to be accompanied by investigation of actual patterns of use. (I assume that many of these claims will be vindicated.)

3. As I was completing the text, the following comments appeared in an editorial in *Nature Biotechnology*: "Herbicide-resistant corn and canola and various pest-resistant crops . . . were what could be produced (because they were technically facile) rather than what the market demanded. They had palpable advantages not to Joe Consumer, but to agricultural producers. . . . What kind of industrial strategist—and we must assume there was strategy at some point—would try to stealthily bring to market what no one needs but everyone has to consume, that the most industry-friendly politician would have difficulty justifying and whose only redeeming feature is to improve the marketing position of the companies that make them?" The editorial's recommendation is that the future of TGs lies in so-called "orphan crops," which would address "unmet needs in narrow markets, such as the specialist nutrition market" (*Nature Biotechnology* 2004).

4. Remember (end of section 6.1) that I am only addressing uses and projected uses of TGs in the growing of major crops. I do not claim that the list presented here is exhaustive. I have compiled it in the course of reading newspaper and magazine articles, articles in scientific journals, and textbooks of biotechnology. It gives a good flavor of what is being promised.

5. Another widely discussed "humanitarian" case is that of the virus-resistant sweet potato, developed by Dr. Florence Wambugu (with assistance from Monsanto) in Kenya. For a detailed critical analysis of this case (that in many ways parallels the discussion of golden rice below, see de Grassi (2003).

6. In the development of golden rice, licenses had to obtained for seventy intellectual and technical property rights from thirty-two different companies and universities. All have agreed to give licenses for free for developments for humanitarian, but not commercial, uses. (The dividing line between humanitarian and commercial uses is $10,000 annual income from golden rice.) See Potrykus (2001) for the details on these matters; cf. Borlaug (2000).

7. On some of the mechanisms whereby agribusiness has undermined smallholding farmers in Brazil, see Lutzenberger and Goldsmith (2001).

8. Sometimes, when confronting the morally laden rhetoric of those that insist that they are using science to "help," it's hard not to see written between the lines, "with gritted teeth": Are you doubting our intentions? Do you doubt that we have done "good"? But the issue is about the relevance of such intentions to how applications of science actually turn out.

9. On "risks" that may be occasioned by agroecology, from the point of view of the P-side, see section 10.5.

10. Compare MST leader João Pedro Stédile: "It was not enough to open the fences and get land. We had to open the gates of ignorance, of capital. And, two years ago, we discovered that we had to open one more: that of technology of multinationals that imposes genetically modified seeds (GMOs) on us. If we lose our seed heritage, it will not matter that we have conquered land and capital. Here in Brazil, the small farmer must fight against GMOs" (Stédile 2004). Opposition to TGs has become a central MST objective, along with matters connected with the environment and land reform; for example, Gilmar Mauro, another MST leader said: "We emphasize the importance of discussing such questions as the environment, biodiversity, the importance of combating genetically modified organisms and of acting against the large agro-industrial monopolies in our country; and we continue to engage in acts of solidarity aimed at propagating a project of agrarian reform that will serve the whole of society" (interview in *Correio da Citadania*, August 19–26, 2000, my translation).

Chapter Nine

Environmental Risks of the Development and Use of Transgenics

P_3 There are no hazards to the environment arising from the current and anticipated uses of transgenic crops and their products that pose risks—of seriousness, magnitude, and probability of occurrence sufficient to cancel the alleged value of their benefits—that cannot be adequately managed under responsibly designed regulations.

C_3 This claim about risks is not well established scientifically. Moreover, the greatest risks may not be direct ones the environment mediated by biological mechanisms, but those occasioned by the socioeconomic context of the research and development of transgenics and their associated mechanisms, such as designating that transgenic seeds are objects to which intellectual property rights may be granted.

The demonstrated efficacy of techno-scientific innovations does not suffice to legitimate their practical implementation. In the case of TGs, it suffices neither to underlie the right of farmers to use them and corporations to develop and market them, nor to justify a public policy that encourages their widespread use, especially one that aims to make their use a prioritized component of the farming of the future. Legitimating using TGs involves not only the generation (or expectation) of benefits, but also showing that there are no relevant risks and no better alternatives—hence the important role of P_3 and P_4^1 (under appropriate interpretations) in the argument. According to the P-side, presumptive cases for both of these propositions have been established to the satisfaction of the bulk of the scientific community. The C-side questions the adequacy of the empirical support that has been offered for them, however, and rejects the presumption in their favor. It also has an ethical/political agenda (section 6.3), but that does not mean that its judgments about risks or alternatives lack empirical backing, just as the fact that many on the P-side have links with agribusiness, or claim to value "science" as a major and indispensable source for the solution of such big and urgent problems as hunger and malnutrition, does not per se cast into question their scientific objectivity (Horton 2004). While political, corporate, and value agendas contribute to the apparent intractability of the disputes about TGs and the hostility that tends to mark them, they should not overshadow the importance of questions for empirical investigation that are involved.

Although risk is a matter open to scientific (systematic empirical) investigation, I will argue that the investigation that has been conducted to date provides evidence neither to warrant that P_3 is soundly accepted—that is, accepted in accordance with impartiality—nor to support decisively that risks (or potential risks) clearly undermine the legitimacy of using TGs. Then, I will discuss the grounds for endorsing—judging it to be sufficiently confirmed to warrant acting on its basis (see section 3.4; section 9.5 for elaboration)—one or other of P_3 or C_3, which (as stated) are generalizations cutting across large classes of TGs. Standard risk assessments, since they deal with allegations of risk one-by-one, do not provide a basis for sweeping generalizations about the risks of TGs or the absence of them, although they may provide support for the sound acceptance of more specific propositions of the type: "This particular kind of TG, used under certain specified conditions, poses no unmanageable risks" (see below section 9.3). The judgment to endorse (or not to endorse) a proposition involves the interplay of cognitive and social value judgments. In the present case this reflects that the strategies adopted in empirical research on risks, and the standards of "proof" to which its conclusions are held, are implicated in social value judgments, often the same ones that led to different appraisals of the benefits of using TGs (chapter 8). The logic of the interplay of the empirical and the valuative is not always well understood, and this contributes to the (false) perception that here we must be in a realm of endless contestation.

In this chapter, my primary aim is to explore the logic of the interplay of the empirical and the valuative, and related methodological issues, that arise when making the judgments to endorse P_3 or C_3, and to argue that these judgments, in turn, are deeply intertwined with the judgments (respectively) to endorse P_4 or C_4. I will not attempt to survey all the *environmental* risks that have been alleged to arise from using TGs, or (except incidentally) to make definitive judgments about particular alleged risks. Nevertheless, my analysis will suggest that—contrary to the rhetoric of the P-side—insufficient investigation has been conducted (to date) to support that the endorsement of P_3 rests principally on an adequate empirical record. Rather its endorsement draws crucially on holding the modern valuation of control (and, sometimes, on related values of capital and the market) and endorsing many of its presuppositions. This does not mean that the matter has been definitively settled against P_3, for the outcomes of the kinds of research that will be described in this and the next chapter could ultimately vindicate it. Nevertheless, it does at the present time support the precautionary approaches favored by the C-side.

9.1 RISK: GENERAL CONSIDERATIONS

Questions and claims about risks pervade the controversies about TGs and appropriately so, since P_3 is crucial for the legitimation of using TGs. Yet, all sorts of conflicting claims about the risks of using TGs have been made in the public debates, and all parties tend to put forward their claims with an air of certitude. Exactly what is at stake here needs to be sorted out. The relevant issue is not whether there are risks or not. No one (not even the most vociferous proponents) doubts that some uses of TGs would involve risks to health and environment. Thus, no one denies that TGs should not be released for the market unless they have undergone and passed risk assessments (certified by adequate regulatory oversight) for their safety. Any techno-scientific innovation (even engaging in research and development of it) involves risks; so, for that matter, does any human action. The relevant issues are magnitude, seriousness, probability, and manageability of risk.[2]

In order to have good empirical backing to endorse P_3 or C_3, it is necessary to investigate the potential hazards of using each particular variety of TG with the following questions in mind. Does an identified potential hazard pose risks of significant magnitude and probability of occurrence, and, if so, via what mechanisms (physical, biological, human failing, socioeconomic) does it do so? What is the time frame involved? Are the tests undergone on these matters adequate, and are the scientific theories that inform them suitable? How likely are the risks of harmful effects under the conditions (including those mandated by regulations) in which TGs actually are (or are expected to be) used, and how serious is the harm? What empirical procedures and theoretical inputs are needed for assessing these things? Is there evidence that the risks can be contained (or their harmful effects reversed) by appropriate management techniques exercised under the oversight of an adequate regulatory system? What regulations for risk management are in place, and are they adequate in view of the extent (and magnitude) of the risks?

A potential hazard is not the same thing as a serious risk. (This is not always appreciated on the C-side.) Studies have shown that a variety of TG soybean containing a gene from Brazil nuts can cause allergies in those who are allergic to nuts (discussed in Lappé and Bailey 1998), and that pollen from Bt corn is fatal to larvae of monarch butterflies under certain laboratory conditions (Losey, Rayor, and Carter 1999). What do these results tell us? Only, that under some conditions some TGs can have effects like these. Allergies and toxic effects on nontargeted species are potential hazards of using some TGs, and so risk assessment studies should address them. The results do not provide evidence that such effects cannot be contained and managed. The soybean with the Brazil nut genetic material was never put on the market precisely because this hazard was identified. In the butterfly case, given significant differences between laboratory and field conditions, it had to remain for further studies to assess the risk this hazard poses in the fields where Bt corn is grown (see section 9.4).

The C-side denies that, given current regulatory procedures, there is adequate empirical evidence to support that there are no serious unmanageable risks occasioned by using TGs. The risks that it claims have not been adequately dealt with concern human health, the sustainability of the environment and the preservation of its biodiversity, the world's food supply, the survival of the small family farm, and even the integrity of the sciences. It itemizes the alleged potential risks, and bases its allegations on empirically (laboratory or field-trial) demonstrated hazards or on theoretical or social analytic arguments that suggest that mechanisms which could underlie them might be operative. The following quotation provides a familiar list of alleged environmental risks:

> The generation of new pests and weeds; the crossing of TG plants with genetically related species that might lead to worsening of pest problems; harm to untargeted species; the drastic alteration of the dynamic of biotic communities, leading to the loss of valuable genetic resources or to the contamination of native species (introducing into them characteristics from distant relatives or even nonrelated species); adverse effects on ecological processes in agroecosystems; the production of toxic substances because of incomplete degradation of dangerous chemical products codified by the modified genes; and the loss of biodiversity. (Nodari and Guerra 2004: 43–44, my translation)

Other commonly discussed items include contamination of non-TG crops by pollen spread by wind and insects from TG crops (see Mellon and Rissler 2004 for evidence that this risk is more extensive and intrusive than hitherto imagined and that such pollution is already far advanced with crops of maize, soybeans, and canola), creating problems for farmers who select seeds for sowing from their harvested crops (Pollack 2004c), and (in the case of Bt crops) creating bacterial immunity to Bt and thus threatening forms of organic farming (Lappé and Bailey 1998; but, cf. Gould 2003; Zhao et al. 2003; see also appendix: 2, "Further Discussion of Risks").

The P-side normally responds by claiming that the above items refer at most to potential hazards (or, in some cases, to theoretical "speculations"), and that there is no credible empirical ("scientific") evidence for holding that the hazards (demonstrated in the laboratory) occasion serious risks to health and the environment when the TGs are grown in the fields and their products consumed. It also counteralleges that there are serious risks of not adopting TG technologies in agriculture: most notably the risk that not enough nutritious food will be produced to feed the world (P_4) (Prakash 2000; Wambugu 1999); and also risks concerning implementing and maintaining national economic development and trade policies, the ambitions of agribusiness (including to profit from its investments in the research and development of TGs) and the neoliberal project, and the continued development of techno-science (especially in biotechnology and related fields such as molecular biology).

Although the two sides agree that most of the items on the list refer to potential hazards, with the thrust and counterthrust of their various allegations it is easy for them to speak past each other. The P-side insists that the C-side has not produced scientific evidence that the potential hazards constitute serious risks, and the C-side that the P-side has not produced relevant evidence (since it hasn't carried out the relevant research) that they do not. The P-side presumes P_3 until such time as it is rebutted by scientific investigation (with a confidence that lends no urgency to conducting investigation much beyond what has already been conducted). The C-side holds that legitimating the use of TGs would depend on compelling evidence supporting P_3. The "burden of proof" issues, which are coming to the forefront here, will be addressed directly in section 9.4. They are not the only relevant issues. In addition, there are judgments about what should count as a (serious) risk, and these are as influenced by ethical and social value judgments as are those made about what should count as a benefit. Moreover, as is assumed in risk-benefit analysis, the seriousness of a risk is to some extent balanced by the value accorded a benefit: in the light of the foreseeable benefits and of assessment of the likelihood of harmful effects, is taking the risks warranted?

When the C-side says, "The likely harmful side-effects are of such severity and ethical significance, far outweighing any foreseeable gains, that taking the risks occasioned by the use of TGs is not warranted," it is not simply reporting the results of empirical inquiry. At the same time, empirical inquiry (as well as informed theoretical analysis) is indispensable to its case, which minimally needs the itemized list of empirically demonstrated potential hazards. To obtain more compelling evidence, however, it would need studies showing that the potential hazards indeed occasion serious risk in the contexts (growing of crops, consuming their products) where TGs are actually used. Often the C-side lacks this evidence—not in circumstances where the relevant research has been conducted and the evidence failed to materialize, but where the potential magnitude, probability of occurrence, and susceptibility to management of the risks have not been significantly investigated. In these circumstances, the C-side is not adequately rebutted on the ground that there is no available evidence against endorsing P_3, and its responsibility is to pose explicit questions for investigation and to stake its case for continuing to endorse C_3 on the outcomes of this investigation. Similarly, it is the responsibility of the P-side to stake its rebuttal of the C-side on these outcomes (and not simply on the claim that there is currently no available evidence against endorsing P_3).

9.2 THE ROLE OF THE MODERN VALUATION OF CONTROL

Legitimating the widespread use of TGs depends on endorsing, with the support of good arguments, both P_3 (no risks) and P_4 (no alternatives). But discussions of risks (P_3/C_3), and the evidence for or against them, dominate the public controversies, and little attention is given to questions about evidence relating to P_4/C_4, so much so that rarely does the P-side address (even for the sake of rebutting) the evidence that the C-side uses to question P_4 (section 10.1). What is the status of P_4? The P-side seems to treat it as not needing to be defended in the face of empirical evidence that allegedly provides some support for C_4. This is a consequence (I suggest) that follows from proponents of the P-side strongly holding the modern valuation of control and endorsing its presuppositions, and thence endorsing P_1 (privilege to research conducted under materialist strategies, and techno-science as principal source of solutions to the world's great problems). To concede that C_4 is a serious competitor open to scientific investigation would be to question P_1. Where the modern valuation of control is held, it tends to be taken for granted that the only viable large-scale alternatives (as distinct perhaps from special niches that may be available to, e.g., organic farming) are conventional high-intensity and TG-oriented farming, these being the forms of farming that are highly informed by scientific knowledge gained under materialist strategies. Then, it is relevant to consider arguments (and the empirical evidence that informs them) that conventional farming might suffice for future productive needs, so that TG-oriented farming would be unnecessary, but once those arguments have been rebutted (and there appears to be general agreement that they have been), the evidence is taken to support P_4 clearly. Endorsing P_4 permits many roles to conventional farming, as well as that individual farmers might make different cost-benefit analyses when deciding how to farm, but the risk assessments that will be relevant to the cost-benefit analyses will be those that compare the potential risks of TG-oriented and conventional farming methods.

Holding the modern valuation of control provides (I suggest) the principal source of the prima facie legitimacy that proponents often feel entitled to claim for current and future uses of TGs (section 9.4 below). As we have just seen, it functions to keep C_4 (and its scientific credentials) from gaining a fair hearing. It also serves to back the claim that a presumption should be accorded in favor of P_3, and thus put the "burden of proof" on the C-side. It does this by virtue of endorsing Presupposition (b) (section 1.1), which affirms that techno-scientific solutions can be found for problems, occasioned by the "side effects"—for example, health and environmental problems—of techno-scientific implementations themselves, by further developments of techno-science.[3] Indeed, TG technology itself is often portrayed as such a novel technology, one that can alleviate the environmental damage caused by an older technology (chemical-intensive farming), since it will supposedly, for example, involve the use of smaller quantities of herbicide or pesticide (section 8.1: Coyne 2003; Rauch 2003). Holding the modern valuation of control, and thus endorsing its presuppositions (even more so, when reinforced by holding the values of capital and the market) tend to nurture a casual attitude towards potential risks. (If using TGs occasions risks, we can be confident that yet new techno-scientific advances will enable us to deal with them adequately; so we don't need to be too cautious in the face of potential risks!) It also nurtures a ready acceptance that, prima facie, a techno-scientific innovation is the bearer of great benefits, to be foregone only if the risks involved are very high indeed. To the P-side, TGs represent a novel power for controlling natural objects, another in the long line of triumphs of modern scientific research, a new opportunity for further embodying the modern valuation of control and deploying scientific knowledge for the solving of great human problems. To question the legitimacy and value of using TGs is, then, tantamount to questioning more generally the value of techno-science (modern science and its technological applications). If using TGs is de-legitimated, what other potential techno-scientific innovations will be challenged? I think that—deep down—this is at the heart of the controversy.

Although the C-side recognizes the novel power and opportunity engendered by TGs and their source in scientific developments, it contests the modern valuation of control (section 6.3). It interprets the positive rhetoric of the P-side, notwithstanding the sincerity of those who use it, as cover for the subordination of the well-being of poor peoples to the interests of those (corporations, the global market, the military, investors) who gain from the further entrenchment of the modern valuation of control (section 6.3; Cayford 2003). Specifically, the C-side doubts that TGs offer much of significance for dealing with problems of the poor, like hunger and malnutrition, and that probably their spread will make things worse,[4] in part because it maintains that their further implementation (even of the celebrated varieties like "golden rice"), and the conditions required for it, reflect the interests of the global market system, the very same system within which poverty, seen as the fundamental cause of hunger and malnutrition, persists today (section 8.3). The C-side contests the modern valuation of control in the name of the values of "popular participation" (section 6.3); it also contests P_1 (chapter 7) and seeks to implement more fully agricultural methods, such as agroecology, which cannot be adequately informed by scientific knowledge principally gained under materialist strategies. Research on the underlying structures, processes, and interactions of phenomena (e.g., seeds and crop plants), and the laws governing them, is able to provide knowledge and evidence about the technical possibilities (and efficacy) of genetic engineering, but it dissociates these phenomena from their agroecological locations and so is unable to address the full range of possibilities open to seeds and crops and the effects (including risks) of their uses on the environment, people, and social arrangements. Knowledge gained under agroecological strategies is pertinent not only to inform farming practices, but also to the appraisal of the side effects (risks) of the use of TGs.

The C-side, I emphasize, proposes a systematic alternative program in agriculture (part of a package of proposals that also pertains to other areas of social life), not in the first instance because of worries about the risks of TGs or because it defends the status quo, but because it wishes to further the values of popular participation, which underlie a different vision of interaction of human beings with one another and with nature (section 6.3). Its objections are first against the modern valuation of control and (perhaps more fundamentally) against many of the values (related to property and the market) of the institutions that nourish them. It does not oppose technologies, which enable novel forms of control, in principle (Guerra et al. 1998a; 1998b; Costa Gomes and Rosenstein 2000); rather it appraises them in relation to how they may contribute to further the values of popular participation and so, in practice, (perfectly coherently) it utilizes some of them. Many (perhaps all) of its objections to the use of TGs are also objections (and are explicitly recognized as such) to using methods of high-intensity farming. For the C-side, TGs represent the latest innovations in high-intensity, industrial agriculture (whose driving forces are the institutions of global neoliberalism), innovations that may occasion some novel risks in addition to those occasioned by high-intensity farming. Thus, for it, the relevant question is not about whether TG-oriented agriculture poses less serious risks than conventional agriculture, but (provided that a good case can be made for endorsing C_4) whether it poses less serious risks than alternatives such as agroecology. All risk analysis involves the comparison of risks posed by different approaches. What should be taken as the relevant approach for comparison depends (in part) on judgments made about P_4/C_4.

9.3 DISPUTES ABOUT RISKS

Although some critics of TGs maintain that there is compelling evidence against P_3, I attribute to the C-side the weaker claim, C_3, whose first sentence may be put:

C_3a: P_3 is not sufficiently well supported by available empirical evidence.

The P-side, when it endorses P_3, maintains (citing the authority of science) that it is well supported by available evidence or, more cautiously, that in the light of available evidence there is a presumption in its favor. In addition, it sometimes suggests that the C-side has no genuine interest in the evidence and is driven solely by its political agendas (appendix: 1) or by "anti-biotech activists, who have nothing but fear-mongering and pseudo-science to support their demands" (Prakash 2000). At the same, however, it often overstates the available evidential support for P_3. When it does this, it suggests to the C-side that it has rushed hastily to claim legitimacy—perhaps because of its links with business interests whose investments might be harmed by delayed innovations of TGs, or because furthering these innovations will provide an impetus for further scientific research and new possibilities for funding favored kinds of research. There is plenty of opportunity here to tar one's opponents with ideological or self-serving labels. Nevertheless, this should not obscure that (as we shall see) there remains a deep question about what counts as sufficient and adequate evidence for a proposition like P_3.

Normally we should not expect to be able to obtain evidence for P_3 that is as compelling as the evidence for, to give an example: "A gene from the bacterium Bt may be engineered into the genomes of maize plants so that they produce a toxin lethal to many varieties of insects." Evidence for the efficacy of TG technology is more clear-cut than evidence for the presuppositions of the legitimacy of using it, and (normally) the same research that produces the evidence for the former is (comparatively) silent on the latter, for the research (conducted under a variety of materialist strategies) that underlies the efficacy of TG technology dissociates from the ecological (and social) context and consequences of its use. Yet, it is not reasonable to withhold endorsement to P_3, and thus to deny legitimacy to agricultural innovations that are informed by it, until we gain comparably compelling and clear-cut evidence for it.

What kinds of research are pertinent to appraising P_3? What counts as "sufficiently well supported by the available and adequate evidence"? Each of the P- and the C-sides should be prepared to give explicit answers to these two questions (and to present arguments for its own answer and against the other's); to lay out the evidence it offers for its endorsement (P_3 or C_3); and to specify what investigation the opposing side would have to conduct (and what sort of outcomes it would have to have) in order to challenge the endorsement it makes. (This is consistent with the interplay of cognitive and social value judgments that occurs when propositions are endorsed.) Certainly, unless it meets these conditions, the P-side cannot properly claim the authority of science for its endorsement of P_3.

When critics endorse C_3a, they may be affirming one or more (perhaps all) of the following items, (i)–(iii):[5]

C_3a (i) Insufficient research has been conducted to identify possible hazards of TGs. (ii) Insufficient research has been conducted on the severity, magnitude, and probability of risk occasioned by the known hazards. (iii) There is insufficient evidence that the risks can be managed so as to avoid serious harm to the environment; or, under prevailing circumstances, it is unlikely that adequate regulatory systems will be implemented.

Elaborating C_3a in this way brings out sharply the complexities involved in appraising P_3. In the first place, terms like "severity" and "serious harm" require interpretation and, contextually, some kind of operational definition. How they are interpreted will necessarily reflect judgments made about expected benefits (a risk to the environment will be judged less severe if it is balanced by environmental or other benefits), and comparisons to the risks occasioned by other agricultural practices. Appraisals of P_2, P_3, and P_4 are deeply interconnected. Relative to chemical-intensive conventional farming, using TGs may occasion less risk, but relative to agroecological approaches, much more; then assessing the risk will be linked with issues about the productive and other possibilities of agroecology and, also, about the likelihood of their widespread use. Second, especially in the light of (iii), it is clear that the matter cannot be settled in the course of research that is limited to the methods of biotechnology (molecular biology, physiology, etc.) and ecology. There is an irreducible social dimension to the appraisal, and risk assessment procedures that abstract from it cannot be decisive. Third, when stated in this way, it is still incumbent on the critics to give reasons why they claim that there is insufficient evidence on these matters. In what way is the evidence that is cited by the proponents insufficient? And what further research projects would they design to get at their reservations?

A Modified Version of "No Serious Risk"

According to the P-side, since hazards may vary with the TG and the environment of planting, the legitimacy of planting a particular TG crop (e.g., a variety of *Bt* maize) does not depend on refuting C_3a (i)-(iii), for it does not depend on the general and open-ended legitimacy of TG technology.[6] Some hazards may pose serious risks in some environments and others will not, and the risks of some but not others may be able to be managed. Risk analysis is (and should be) carried out *case-by-case*, TG/environment-by-TG/environment. Then, it would be better to attribute to the P-side a narrower proposal than P_3, perhaps along the following lines:

P_3a (i) The TGs that have been released for use in the United States (and countries with comparable regulatory agencies), sometimes with their use confined to certain geographical regions and conditions, introduce no environmental hazards that pose serious risks of significant magnitude and probability of occurrence which cannot be adequately managed under responsibly designed and administered regulations; and (ii) Current regulatory procedures are adequate to identify serious risks concerning new TGs that may be introduced and thus to prevent the release of risky ones (and to contain harm should a risk become apparent after introduction), and these procedures are regularly under review so as to ensure that they continue to be improved.

If the C-side were to acknowledge that P_3a (i) is well supported by the evidence (which it does not, section 9.1), then it would not be able to deny legitimacy to plantings of the currently released TGs on the ground that these particular plants (qua plants growing in farms) pose environmental risks. But, the C-side not only questions (ii) (see below), but also declines to reduce the question of risk to a case-by-case matter, since it considers the currently released TGs to represent the first steps of a trajectory that involves the radical transformation of farming worldwide. It may deny legitimacy to the planting of these TGs on the ground that they are integral (initial) parts of a trajectory that may occasion serious environmental risks. This is not an ad hoc move. After all, the companies that are developing TGs publicize that currently available TGs are "first generation," forerunners of greater things to come (and they would not have been developed in the first place if this were not the case). So, too, do the Foundations and NGOs that want to develop varieties of TGs that they think may serve to meet the needs of poor people in "developing" countries (appendix: 1; section 8.3). Long-term research and development projects have been planned on the basis of this understanding. To the C-side, the legitimacy of plantings of particular TGs is deeply linked with the *legitimacy* of the planned trajectory of TG research and development. Not so to the P-side! Nevertheless, to it, the *value* of these crops is linked with the expected legitimacy (as well as perceived desirability) of the trajectory, so that I interpret it to add the following item into P_3a:

(iii) TG technology (regulated under the procedures mentioned in (ii)) may be safely expanded (so far as environmental impact is concerned) even to the point, as the current trajectory of its development may suggest, that the practices it informs come to be a major and widespread component of farming from now on (or at least to have much greater salience than they currently do).

9.4 BURDEN OF PROOF

Do the risk assessments that have been conducted under the best current regulatory procedures provide sufficient evidence for P_3a, or good grounds to grant a presumption in its favor, so that the "burden of proof" should rationally fall squarely on the C-side?[7] The P-side claims that they do. We saw (section 9.2) that holding the modern valuation of control and endorsing its presuppositions tend to support a presumption in favor of P_3. More importantly, however, the P-side claims that a sufficient number of risk assessments have been conducted to ground confidence in P_3a. Thence, it puts the burden of proof (to rebut P_3a) on the C-side and (in the context of countries with well-functioning regulatory bodies) this is, and is intended to be, a difficult burden to assume. Given (ii), its successful assumption would presumably be to identify a hazard that, according to standard risk assessment procedures, poses a serious risk for which there is evidence that it is unlikely to be contained given current (and foreseeable) regulatory procedures.[8]

The burden arises for the C-side, however, only in the light of the P-side having carried (as it claims it has) the "prior burden," that is, showing that the released TGs have passed its risk assessments and that there is regulatory oversight in place to ensure that no releases of TGs occur without their having passed properly designed and implemented risk assessments. Unless the prior burden is carried, the defense of P_3 that appeals to "there is no scientific evidence against P_3," would be empty, for there would also not be any evidence in favor of it; then the absence of counterevidence might simply indicate the absence of research on the matter. Thus, the importance of carrying the prior burden should be emphasized, and that continuing to carry it requires that conducting a complex body of research projects remain an ongoing concern. The outcomes of risk assessment can vary in critical ways with the kind (and specific variety) of TG and the environment of planting. For example, "no risk" in one environment does not imply "no risk" in another; TGs released for planting in the United States, having passed risk assessments deemed adequate by U.S. regulatory agencies, may yet pose serious risks in another agricultural environment (NRC 2002: 247) That is why my formulation of P_3a is so heavily qualified.

The prior burden cannot be carried, once and for all, under one set of conditions as a stand-in for all conditions. This is a fundamental principle of risk assessment, not a demagogic assertion invented by more radical adherents of the C-side. While sound judgments of the efficacy of TG technology travel easily across national and regional boundaries, the outcomes of risk assessments do not. (For a striking example of how this point can easily be misunderstood, see appendix: 2, concerning legal and legislative actions in Brazil.) Have those who endorse that all the TG varieties that have been released in the United States can be presumed to be safe adequately assumed the "prior burden"?

Monarch Butterfly Case

The celebrated case of the monarch butterfly illustrates why proponents are convinced, but also why critics are not. A study published in 1999 showed that pollen obtained from Bt maize grown commercially (and fully in accord with regulations) is toxic to the larvae of monarch butterflies under certain laboratory conditions (Losey, Rayor, and Carter 1999). Thus, it identified a hazard that had not been investigated in the risk assessments conducted prior to approving these crops for use. Subsequently, risk assessments were conducted. The investigators concluded that this hazard did not pose a serious risk to the population of monarch butterflies in the farm fields in which Bt maize is grown, except perhaps in the case of one variety of Bt maize, which (they noted) was being phased out from the market (Sears et al. 2001; Hellmich et al. 2001; Oberhauser et al. 2001; Pleasants et al. 2001; Stanley-Horn et al. 2001; Zangerl et al. 2001). According to the investigators, the toxicity observed in the laboratory was apparently an artifact of the quantity of pollen to which the larvae were exposed, and also, contextually, it is irrelevant, since in the farm fields the larval stage usually has passed before the maize plants produce their pollen. This outcome has been interpreted by P-side as further empirical support for the environmental safety of TGs, and the well functioning of the regulatory system: no harm was done by crops that had been approved under the regulatory measures; and, when a new hazard was subsequently identified, the regulatory authorities responded appropriately with post hoc risk assessments to check whether or not it posed a serious risk.

The C-side interprets the matter differently. First, it points out that the potential hazard to the population of monarch butterflies was not identified prior to approval of these crops (and that one variety, which may have had some, clearly not catastrophic, impact on the larvae, was used for several years). Second, it claims, this suggests possible weaknesses in the hazard identification processes, and raises the question—one of great salience for risk assessments of Bt crops planted in other countries—of what other nontargeted insects, which have not been identified, might be affected adversely by Bt crops. More generally, it raises the concern that the hazard identification processes might involve at their core mere guesswork, or (perhaps) carelessness or very casual induction based on earlier studies. Third, when the hazard was identified, there were no procedures in place to deal with investigating it immediately; the response was ad hoc (special funding had to be arranged and proposals to carry out the investigation solicited), reflecting that there are no normally functioning mechanisms for ongoing scrutiny for risks that might only become apparent post hoc. Fourth, the C-side adds, the recourse to an ad hoc response and the expense involved in carrying out the post hoc risk assessments suggest that it is unlikely that thorough comprehensive risk assessments can be expected generally to be made, especially as the quantity, variety, and complexity of TGs increases (see NRC 2002: 192ff, on the need for "post commercialization testing").

Has the "Prior Burden" Been Carried?

The C-side tends to maintain that the process of risk assessment is not only marked by guesswork, lacking a systematic means for identifying hazards and dependent on ad hoc response mechanisms, but also the procedures for enforcing regulations are inadequate and frequently flouted (Pollack 2003; 2004a; Clarke 2003). Moreover, it points out that the results of risk assessment studies may not be trustworthy, for often (not always) the studies are conducted and their results held confidential by the corporations, who conduct most of them, who have an interest in positive outcomes,[9] and who—by exercising the rights (IPR) they hold to transgenes and seeds—may prevent safety studies from being conducted (see Dalton and Diego 2002).

In addition, the criticism that the process of risk assessment is insufficiently attentive to known hazards has recently gained attention (Powell 2003; Mellon and Rissler 2004; Pollack 2004c; NYT 2004b). A well-known hazard is that TGs may contaminate farm crops and related plants in the environment. Pollen flow (with the movement of winds, insects, etc.) is the principal mechanism of contamination. Since pollen flow, and resulting crossings of plants, is a fundamental fact of nature, it is not obvious that contamination by TGs poses a serious risk, and there is no evidence to date that it has caused significant environmental harm (Pollack 2004c). But, recent studies show that the extent and rapidity of contamination expected with the release of TGs have been considerably underestimated. Mellon and Rissler (2004) show that, "Seeds of traditional varieties of corn, soybeans and canola are pervasively contaminated with low levels of DNA sequences obtained from transgenic varieties." Underestimation of the extent and rapidity of potential pollution is common. It was estimated by its developers that the pollen of a variety of TG grass (a creeping bentgrass, resistant to RoundUp), being developed (by Monsanto and Scott Corporations) for use on golf courses and home lawns, would travel for only about one thousand feet, "but when Environmental Protection Agency scientists studied gene dispersal from some 400 acres of genetically modified grass, they found that some genes reached sentinel plants of the same species as far as 13 miles away and wild relatives almost 9 miles away" (NYT 2004b). In this case, the regulatory body is attentive to the potential hazard, and it has stayed commercial release of the grass pending a thorough environmental impact study. At the same time, however, it raises questions about the adequacy of the environmental impact studies that have been conducted on the TGs that have already been released. *The New York Times* concluded: "Whatever they decide about bentgrass, regulators will need to reassess whether they are looking hard enough and far enough for the potential impacts of genetically modified plants."

The C-side, in view of the above discussion, denies the presumption in favor of P_3a. Since (it claims) the prior burden has been assumed only in a perfunctory way, it rejects that it is reasonable to expect it to pick up the burden of proof until such time as the prior burden has been clearly and forthrightly carried.

To the P-side, their opponents' criticisms are overstated, at most pointing to the need to improve the regulation of risks and the conduct of risk assessments, but to nothing that cannot easily be addressed.[10] They are also misguided, since they ignore not only the already demonstrated and currently anticipated benefits of TGs, but also P_4. Why, the P-side asks, should steps beyond standard risk assessments be taken to deal with hazards (identified or as yet unspecified) that have not been demonstrated to occasion serious risks,[11] when there are so many clear benefits to be gained (and much to be lost by those who have invested in developing TGs), when the kind of risks that might arise are of the same kinds as those that might arise within conventional farming, and when there is no serious alternative approach? The case now for the presumption in favor of P_3a, leaving aside the background of the modern valuation of control, rests on the risk-benefit analysis that the potential harms of using TGs (so far we can anticipate them) are more than cancelled out by the value of the benefits to be gained. Clearly evidence that supports endorsing P_4 would strengthen the presumption; if there are no relevant alternatives, tolerance of risks should be greater. But (anticipating section 10.1 and as already foreshadowed in the presentation of agroecology in section 5.4), adequate evidence of this kind is not available at present. In this context, the presumption must be based on the value of the benefits and, especially since these benefits are not uniformly valued across value-outlooks (chapter 8), on having adequately carried the prior burden that, among other things, should show that there is a regulatory and management system in place that can take care of any risks that might come up.

Here the issue of contamination is instructive. At present (according to the studies cited) the degree of contamination in any locale is quite small, but it has spread widely to many locales, although no actual harm to ecosystems (as distinct from agroecosystems—see next section) has been documented. However, although we do not know how rapidly these small effects will multiply and accumulate, and how harmful their effects might be on ecosystems and on how wide an area they will have impact, there can be no doubt that they will accumulate and have impact over a wider area. We can only "wait and see" how harmful the effects might be and over what temporal and geographical scale, but whatever the harm (if any) eventually incurred, it is likely to be permanent and irreversible. For the C-side, the acclaimed benefits of using TGs are not so great as to warrant pursuing them, without the prior burden having been clearly carried, in the face of potential great harms that would irreversible. It is not enough that currently performed risk assessments and subsequent monitoring of TG crops and their products have not demonstrated that there has been any significant short-term harm to ecosystems and human health. To act informed by the endorsement of P_3, citing in support of doing so that there are no demonstrated short-term harmful effects, is effectively to postpone questions of the legitimacy of using TGs on a large scale until it has become a fait accompli (when it will be too late to act on the basis of negative answers). That is not a reasonable proposal; and when it is integral to corporate practices or government policies, one should not be surprised if opponents conclude they have no alternative but to resort to direct action (e.g., burning fields planted with TG crops, destroying TG products) in an attempt to bring their objections to the public.

I have identified two unreasonable extreme stances: first (in section 9.3): to withhold endorsement to P_3, and thus to deny legitimacy to agricultural innovations that are informed by it, until the evidence enables us to affirm with confidence that it has been soundly accepted in accordance with impartiality; and (now): to make a presumption in favor of endorsing P_3, pending the obtaining of definitive evidence of serious harmful effects. They are unreasonable because they hold their opponents to what are, in the circumstances, unachievable "standards of proof." The first stance requires that P_3 be accepted in accord with impartiality, although reasonable endorsement is all that should be expected; the second presumes P_3 on the basis of short-term investigations, although the pertinent challenging evidence refers to long-term effects.

The Precautionary Principle

The middle ground, which I support, expects the P-side to carry the prior burden: to conduct rigorous risk assessments for all TGs in the environments in which they will be used, to be forthright in defending the adequacy of the assessments in the light of criticisms (and to revise their procedures where necessary), and not to release them for commercial use until they have passed these assessments; and to show that there are transparent and enforced oversight procedures in place, framed by well-designed regulations, to ensure that the planting of TGs and the consumption of its products are subjected to ongoing monitoring, adequate to identify (and respond to) any risks that might emerge. The middle ground also expects the C-side to specify concretely (albeit subject to modifications that are justified by appeal to new empirical discoveries) the conditions under which it would accept that the prior burden has been carried. (If the prior burden has not been carried to its satisfaction, it should be prepared to specify what further investigation would have to be conducted to settle the matter at issue.) After the prior burden is carried, the burden of proof would then fall squarely on the C-side.

I do not pretend that it is easy to stake out the middle ground. What is at issue is endorsing or not endorsing P_3, so that ethical/social as well as cognitive value judgments are in play. The value-outlooks of the P-side will incline toward a more lax interpretation of carrying the prior burden, and those of the C-side (for the projected risks are more threatening for its interests) towards a sterner interpretation. This is inevitable. To occupy the middle ground is not to anticipate reaching ready agreements, but it is to being committed to make explicit what would count as carrying the prior burden (be it under a lax or a stern interpretation).[12] Explicitness would make it possible for the C-side to identify what it considered too lax about the P-side interpretation (e.g., that it has not taken into account the geographical scale of potential contamination, or that too much of data from risk assessments is held confidential by agribusiness corporations); and conversely for the P-side to identify what is too stern about the C-side interpretation (e.g., that it ignores the costs of risks assessments, or expects that potential hazards, which are proposed for consideration only on speculative theoretical grounds, should be investigated rigorously before release of the relevant TG). This would make possible focused argument, again not necessarily agreement, but articulating the viewpoints in a way that could inform discussions in a democratically constituted forum, in which a practical resolution might be sought for adoption in public policy (which, of course, will vary with different conditions in different countries—see appendix: 2, "Public Policy: National Variations"). The middle ground pays attention to the value-outlooks of both sides and takes into account both the benefits (valued given its outlook) claimed by the P-side and the potential harms (of particular concern given its outlook) noted by the C-side, and attempts to characterize how empirical investigations can (and should) be taken into account in reaching a democratically acceptable compromise—at least until such time (if there ever is such a time) when there will be decisive evidence in favor of one side or the other.[13]

The argument just made for staking out the middle ground underlies adopting (what has been called) the "Precautionary Principle." The basic idea of this principle may be put as follows: "It is legitimate for a country to prohibit the use or import of a technology (and its products) for the sake of gaining time to investigate risks (to health or environment), even if there is not definitive scientific evidence available (prior to further investigation of alleged risks) that there really are risks" (see appendix: 2). It puts the burden of proof (what I have called the "prior burden") on the producer of a new technology to show there are no significant risks occasioned by using it, but in a way that the burden is considered to be carried if relevant investigations are to have favorable outcomes: the burden is carried if all plausible potential hazards (i.e., hazards that have been demonstrated in the laboratory or identified in the field, or that have a sound theoretical foundation) have been dealt with adequately. The precautionary principle does not license the ongoing and open-ended appeal to unknown risks to de-legitimate a techno-scientific innovation. The rationale for adopting it is strengthened when we also take into account that the socioeconomic context of the use of TGs provides further mechanisms for risks, social mechanisms for environmental risks as well as for social risks.

9.5 RISKS THAT DERIVE FROM THE SOCIOECONOMIC CONTEXT OF USE

The adoption of agroecological strategies is relevant not only to the investigation of agricultural alternatives (chapter 10), but also to the investigation of risks, for crop plants and their seeds, including TGs, are objects of agroecosystems—agricultural-ecosystems; ecosystems that are integral parts of agricultural spaces, that is, of a type of sociocultural-economic environment—and it is an abstraction to consider them simply as biological entities (section 7.2). The "sustainability" (section 5.4) of TG-oriented agriculture, for example, is an agroecological issue, and investigation of it should not dissociate from the social and economic relations of production of crops and the distribution of their products (and the causal networks of which they are a part). When we attend to what TGs are without abstraction, P_3 becomes interpreted as:

P_3b There are no environmental hazards arising from the current and anticipated uses of TG crops—in the agroecological systems in which they will actually be used—and their products that pose risks (of seriousness, magnitude, and probability of occurrence sufficient to cancel the alleged value of their benefits) that cannot be adequately managed under responsibly designed and administered regulations.

TGs have effects (on health, the environment, social relations), not only qua biological entities (entities studied in the biological sciences), but also qua objects of agroecosystems, which have irreducibly social dimensions (section 7.2), including (at least for the TGs available and foreseeable today) that the interests of capital and the market shape the relations of production and distribution (section 8.3). I pointed out in the previous section that the C-side denies that there is a presumption in favor of P_3a. Usually, however, it goes further, and denies that P_3a suffices to play its intended role in legitimating the widespread use of TGs and claims that only P_3b could suffice. Those who appeal to P_3a in arguments for the legitimacy of using TGs usually interpret "environmental hazard" as potential harm that is directly due to planting, tending, and harvesting the crops, where "directly" normally means explicable in terms of molecular, genetic, and cell biological, plant physiological and "natural" ecological mechanisms (NRC 2002: 56 ff). Thence, standard risk assessments can only take into account effects of TG seeds and crops, qua biological entities, but not qua members of agroecosystems. (They do not, e.g., consider the risks of TGs, qua the intellectual property of large agribusiness corporations.) These risk assessments are incomplete in two ways: on the one hand, they abstract from potential ecological harm that may arise from the socioeconomic context of the research/development/use of TGs; and, on the other hand, they abstract ecology from social ecology by not addressing potential social harms, for example, those that might arise from corporate control of the food supply that is furthered through the granting of intellectual property rights for TG plants and related engineering procedures.

The C-side is wary of any judgments made where these two abstractions are not challenged. That is why it maintains that P_3 needs to be interpreted as P_3b in order to play its role in arguments for the legitimations of TGs, and also why it is unimpressed by an argument that has recently gained currency: "Once P_3a has been endorsed (following standard risk assessments), farmers have the right to use TGs (the ones that have passed the risk assessments) if they choose to do so." Wariness in the face of these abstractions, however, following the logic behind the middle ground, should be expressed explicitly and not be turned simply into the demand for more evidence for P_3a. That would be obfuscating what is at issue for, to the C-side, it is not only "more evidence" (more of the same kind of evidence), but also (and principally) "different kinds of evidence" that is at issue. Standard risk assessments, no matter how thoroughly they are carried out, do not address either environmental or social risks that may be occasioned by socioeconomic mechanisms. A compelling case for P_3a would not amount to one for P_3b,[14] for (the C-side claims) there are good reasons to think that serious risks are involved, some of which are connected with the socioeconomic relations of production of using TGs, that will only become apparent in the long term—potential harmful effects to the environment, to the maintenance of biodiversity, and to the preservation, regeneration, and creation of sustainable, productive agroecosystems. There are also risks to the livelihoods of poor farmers, to the projects of agrarian reform that they favor, and to their ability to be able to farm in ways that express their own values. It should be noted that, when considering evidence related to P_3a, it is not always possible to separate the narrow notion of environmental hazard from a social one, for judgments about what constitutes a serious environmental hazard often depend on the social context. For example, one environmental hazard of Bt crops is that resistance to the Bt toxin will be developed in the targeted pests. How serious one considers the risks posed by this hazard depends (in part) upon the value that one accords to the organic farming practices that will be disrupted if (or when) those pests become resistant. (Discussions in NRC 2002: 236ff and Thompson 1997; 2003a address this kind of complexity.) The contamination effects of growing TGs raise similar issues.

Contamination of Mexican Maize Landraces

There are well-known environmental mechanisms (pollen spread by wind and insects) that ensure that the genes engineered into TGs will spread, to some degree or other, to crops in nearby fields and into related plants in the environment. The C-side calls this phenomenon "contamination," implying that it represents a hazard and, clearly, if its effects are concentrated, it is one that might pose a risk to farmers who wish to grow non-TG crops, who may lose "non-TG" certification because of the contamination or, if they save their own seeds for replanting for the next year's harvest, they might be sued for violating the patent rights of the corporation who sells the seeds. Nevertheless, there is dispute about how serious and unmanageable are the risks occasioned by contamination. Above I briefly discussed recent studies that show that the rapidity and extent of contamination is much greater than had been surmised in the light of earlier risk assessments, and until more is known about these issues a confident judgment about the risks posed by contamination cannot begin to be made. Then, questions about seriousness of risk are likely to have a social dimension, and socioeconomically generated mechanisms may also be implicated in the spread of the contamination with potentially far-reaching ramifications (Tokar 2004).

Recently evidence has been offered that TG genetic materials have contaminated Mexican maize landraces. (Some regions of Mexico provide the world's major center of biodiversity of maize). An important part of the mechanism of contamination is that peasants have planted U.S.-produced maize (which happens to be TG) that they were able to buy cheaply (supposedly for processing and consumption) in part because of U.S. farm subsidies and provisions of the NAFTA agreement (Santis 2004). Then, ordinary mechanisms of gene flow across environments led to the TG DNA becoming part of the genomes of some of the local varieties. This example has occasioned considerable (and bitter) controversy. The first published study offering evidence for the contamination (Quist and Chapela 2001) was severely criticized for alleged methodological shortcomings (Metz and Fütterer 2002; Kaplinsky et al. 2002; Worthy, Strohman, and Billings 2002; Hodgson 2002a; Martinez-Soriano et al. 2002; Christou 2002). Even the editors of *Nature* (*Nature* 2002) wrote that they had made a mistake in publishing the article, but Quist and Chapela stood by their basic claims (Quist and Chapela 2002). Since then the fact that this contamination has occurred has been confirmed in studies made by Mexican scientific bodies (see Schapiro 2002; also ETC Group 2002).

There remains the question of the significance of the fact, and some on the P-side insist that it has no more significance than the fact that the genetic constitution of landraces is constantly changing. The C-side points out that the only certainty here is that any introgression of TG DNA into the landraces will be irreversible and that it has unforeseeable consequences for the future. This is a matter of considerable concern since cross-breeding with landraces from the centers of biodiversity is part of the process of generating new crop varieties and replenishing seed supplies, and the stewardship of these landraces has been one of the great cultural achievements (and gifts to humankind) of the Mexican peasantry. Since the above-mentioned mechanism of the contamination (documented in Schapiro 2002; Santis 2004) demonstrates the absence of effective regulatory oversight, the concerns are even more serious. (It is illegal to grow TG maize in Mexico, but the peasants who planted it had bought the maize on the market principally for their families' consumption, and they used the occasion (since it was cheaper than buying needs) to select out some of the seeds for planting. In these circumstances, that the seed was TG, and that therefore planting it was illegal, was probably not even known by them.) The agroecosystems in which TG maize is produced and distributed include the practices of international trade and humanitarian aid (and subsidies to farmers in the United States which enable the production of a vast surplus of some crops for export and foreign aid); in them, farmers in "developing" countries may plant seeds they buy or receive as food aid in a context where there are no regulatory oversight mechanisms effectively in place (regardless of decreed regulations that do not permit growing such seeds) to curtail undesired consequences.[15] Of course, if these features of the agroecosystems of TGs were to change, it might turn out that P_3b could be soundly accepted. That possibility is not enough to legitimate the emphasis on TGs now:

> Heedlessly allowing the contamination of traditional plant varieties with genetically engineered sequences amounts to a huge wager on our ability to understand a complicated technology that manipulates life at the most elemental level. Unless some part of our seed supply is preserved free of genetically engineered sequences, our ability to change course if genetic engineering goes awry will be severely hampered. (Mellon and Rissler 2004: 2)

Potential Social Harms

Other kinds of risks are occasioned, too, when we consider TGs, qua objects in agroecosystems in which the interests of capital and the market (represented by agribusiness) shape the relations of agricultural production and distribution—in particular, risks to the development, or even continued existence, of alternative forms of farming. The widespread and rapidly expanding use and rapid development of TG-oriented farming, that the P-side supports, are incompatible with (or at least seriously hinder) the significant development of agroecological farming. First, since key components of TG technology are bearers of IPR held mainly by large agribusiness corporations, TGs are instruments of profit and of increasing the control exercised by agribusiness over virtually all dimensions of the agroecosystems in which they are used. They cannot have significant uses in agroecology, which aims to further such values as local empowerment (section 5.4). Second, agricultural spaces in which agroecology is practiced are likely to be reduced because of the expansionary objectives of the P-side, furthered by its economic power and sometimes by government or international policies that give priority to TG-oriented farming. Third, where TG-oriented farming is prioritized, and P_4 endorsed, there is little support (including funding cuts—Dalton 2003; Knight 2003) for engaging in research under agroecological and related strategies, so that knowledge needed for improving agroecological practices will not be obtained. Moreover, it is likely to lead to the downplaying (and effective loss) of time-tested methods and the knowledge that informs them (see discussion of Zhu and Wolfe in section 10.1 and Shiva in section 7.2). Fourth, it may also threaten the conditions upon which alternative forms of farming depend. These conditions include the availability of a market for products and the strengthening of biodiversity. The former may be threatened by the conduct of trade—massive farm subsidies enable the export of cheap U.S. products that undersell the products of poor producers in many "developing" countries (NYT 2003a). In addition, since the efficient use of TGs normally requires plantings of monocultures, their spread and the accompanying greater control of agroecosystems gained by agribusiness threaten to undermine maintaining the biodiversity (nourished by agroecology) that is needed to develop and replenish crop seeds, as well as biodiversity more generally in the environs of agroecosystems. Relatedly, it may also undermine effective methods that have been developed by scientific research with ecological, rather than techno-scientific emphases. For example, in Brazil large increases in the productivity of soybeans have been obtained by using the bacterium *Rhizobium* to improve the nitrogen-fixing properties of soybean plants in deficient soils, but intensive use of glyfosate (the active ingredient of RoundUp) inhibits the metabolic action of Rhizobium (Albergoni, Pelaez, and Guerra 2004).

The third aspect also means (section 10.4) that research that needs to be conducted for the sake of empirically appraising P_4 (and, therefore, P_3) will not be conducted. The upshot of all four aspects combined is that, because of these features of the agroecosystems in which TGs are used, P_4 may *become* true in the future; agroecology will not be an alternative, not because its potential was investigated and shown to be severely limited, but because space and knowledge were denied for its development. This is an enormous social risk, for (according to the C-side) agroecology offers potential benefits to small-scale farmers and their communities in "developing countries" in a way that is unmatched by developments of TGs and high-intensity farming. The curtailment of agroecology could exacerbate rural hunger, displace more people from the land (forcing them to migrate to cities), and eliminate their hopes for agrarian reform. Certainly (section 8.3), research, informed by P_1 and the promise of techno-scientific solutions to problems like hunger, that dissociates from the causal nexus of hunger, including its socioeconomic dimensions, can provide no evidence to the contrary—and, if it is effectively the only kind of research conducted, TG-oriented farming may occasion risks that are not even conceptualized.

Once again, standard risk assessments are not designed to deal with risks like these that arise from the socioeconomic relations of the use of TGs (see section 10.6).[16] For many on the C-side these are the most serious risks, so that until methods are devised to take them into account—methods that involve long-term comparisons of the ecological and social effects of various farming approaches, as well as empirical analyses of the causal nexus of phenomena like hunger and malnutrition—the prior burden will remain not carried by the P-side. When P_3 is interpreted as P_3b, adherence to the precautionary principle gains greater ethical weight. (And once again, if compelling evidence were to be offered for P_4, the issues would have to be recast.)

9.6 THE LOGIC OF ENDORSING PROPOSITIONS ABOUT RISKS

In this section, I will both summarize and elaborate issues about the logic of endorsing one or the other of P_3/C_3 that are implicit in the argument presented above. Disagreements about them reflect fundamental differences in value-outlooks (modern valuation of control-values of capital and the market/values of popular participation) and, because of this, lead to the adoption of different strategies in the empirical investigation pertaining to P_3. Then, the adoption of strategies presupposes answers to the kinds of questions that I posed in section 9.1, for example: What is considered to be a serious risk? What are the mechanisms of risk? What are the evidential standards for judging that unmanageable risk is not present? Who should assume the "burden of proof"? There are no neutral "scientific" answers to these questions. Depending on which value-outlook is held the answers given will be different. This does not mean that scientific research is irrelevant to addressing P_3, only that whatever scientific research is conducted on risks already presupposes answers to these questions, and the strategies adopted in the research will be consonant with these answers. (The research is held to the condition of long-term fruitfulness, so there is empirical constraint on continuing to adopt strategies.) All of this fits well with my general account of the dialectical relations that may exist between adopting a strategy and holding a particular value-outlook. Then, one may suggest, the absence of results about risks that are acceptable in accordance with impartiality is a consequence of the relative novelty of TGs, and of not enough time having passed for the complexities to be attended to, so that it is something that may be surpassed with the conduct of further research. Provided that the essential interaction of judgments about P_3 and P_4 is kept in mind, I think that there is a great deal of truth in this suggestion. It needs to be qualified, however, in the light of another source of the difficulty and confusion surrounding inquiry into risks.

Negative Existential Logical Form of "No Serious Risk"

As noted in section 3.3, the very logical form of P_3 (and P_4)—negative existential—poses difficulties for designing appropriate research strategies. Empirical evidence for P_3 (whether interpreted as P_3a or P_3b) is ultimately the absence of empirical evidence against it. Evidence against it would be the actual identification of an environmental risk of significant magnitude and likelihood of occurrence that is intractable to management. But absence of evidence against it is not per se evidence for it; inability at the present time to identify such a risk provides evidence for P_3 only if appropriate and sufficient research has been conducted.[17] Otherwise, not identifying serious risk may indicate only that the relevant and sufficient research has not been conducted. There are two issues here: *relevant* research and *sufficient* research. Is the research relevant to address the serious potential risks, for example, social as well as ecological ones?

Sufficiency raises further considerations, where (once more) value-outlooks make a difference. We have already seen that the P-side affirms that enough research has been done on risks, and that, therefore, the C-side should assume the burden of proof. The C-side denies this, and I argued above (section 9.4) that it should assume the burden only after the P-side has carried a "prior burden" of providing evidence that no specific hazards which have been identified occasion serious risks, where evidence must be provided against specific claims (grounded either in experimental studies, or field observation, theoretical analyses, or socioeconomic studies) that have been made. Carrying that burden successfully depends on appropriate locally available regulatory oversight. Once it is carried, the burden of proof then shifts to the critics. To assume this "posterior burden" requires identifying further specific hazards for risk assessment. It involves identifying further specific research projects to be carried out—more of the same kind of research that led to the judgment that the "prior burden" had been successfully assumed. The logic here is clear enough: Given the logical form of P_3, evidence for it can never be conclusive, and this enables opponents (if sufficiently motivated) always, no matter that evidence has accumulated against specific claims of risks, to keep saying indefinitely, without falling into contradiction, "not enough evidence!" for who knows what the unknown risks may be? Opponents with this kind of stance effectively are putting the matter beyond the purview of scientific research; for them considerations of social values settle the matter without the need for empirical investigation. They do not endorse C_3; they opt out of investigation—and endorsement is a stance adopted following empirical investigation.

What is enough evidence? That depends on how serious are the ethical stakes involved. They are high for both sides, but they point in different directions. The C-side emphasizes potential threats to valued sustainable agroecosystems and to responding to the needs of poor farmers and their communities; the P-side emphasizes potential benefits to the farmers who plant these crops and the corporations that produce them. The values of the C-side side push not only for research under a plurality of strategies that would enable P_3b to be tested, but also for higher standards of testing with respect to P_3a. (In view of the seriousness of the risks that it discerns, taking them needs to be compensated by greater surety that they can be dealt with adequately. Note: higher standards of empirical testing, not that empirical testing is irrelevant.) At the present time the two sides make opposed judgments about P_3a. The matter is not settled. The P-side endorses it, claiming that it is supported by sufficient evidence to legitimate plantings of TG crops. The C-side endorses its negation. Both sides appeal to evidence. But they appeal to different standards (as well as kinds) of evidence, and their different appeals are implicated in their holding conflicting value-outlooks. The judgments made about P_3 thus involve considerations of both cognitive and social values. They are not about whether or not it may be properly accepted in accordance with impartiality (although some partisans on both sides use rhetoric that falsely suggests that they are putting forward empirically settled proposals). They are about whether or not P_3 has sufficient empirical support to warrant its role in legitimating uses of TGs. (They are thoroughly intertwined with further judgments about the value of expected benefits and value judgments that are linked with the context of investigation of P_4.) These judgments are irreducibly both cognitive and social value judgments. They are not judgments of *acceptance* and *rejection*. I have called them (section 3.4) judgment of *endorsement* or *non-endorsement*.

P_3 is not soundly accepted at present. It does not manifest the cognitive values so highly that further investigation is unnecessary, and it could not become accepted without further investigation. But the P-side only needs to make the judgment that it is sufficiently well confirmed to warrant its role in legitimating uses of TGs. Further investigation might lead to the rejection (in accordance with impartiality) of P_3 or to its acceptance (or to no definitive resolution). This has to be taken into account in arguments about legitimacy. Suppose that further investigation does lead to the rejection of P_3. That would mean that its current use in arguments for legitimacy would have led to risks being taken that could not be adequately taken care of (as claimed by P_3). That would not matter much if the risks were not very serious (ethically—given one's value-outlook), but otherwise it would.

So, the greater the ethical seriousness of the risks involved, the greater needs to be our confidence that the evidence supports P_3, or needs to be the degree to which P_3 manifests the cognitive values in the light of available empirical data. If one judges (reasonably) that P_3 manifests the cognitive values to a sufficiently high degree, then one does not have to take into account further the harms that might be brought about should P_3 turn out to be false (rejected in view of further evidence). But different assessments of the ethical significance of the potential harms will lead to different judgments of what is sufficient evidence, and thus different judgments about P_3's role in legitimations of uses of TGs. Whether or not to endorse P_3, in the light of this, cannot be based solely on available empirical data (or its absence) obtained from risk assessments, or even from analysis of risks expanded to consider socioeconomic mechanisms and long-term and large-scale consequences, and the possibilities of alternative methods of farming. That does not mean that data on these matters are unimportant, only that they are indecisive; then, value judgments influence what count as appropriate standards of evidence and what are the most salient data. Commitment to the modern valuation of control inclines one towards endorsing P_3 (or making a presumption in its favor); commitment to the values of popular participation incline one against doing so. Recognizing this makes clear that the dispute about TGs is—to a significant degree—one about social aspirations and the values they reflect and how forms of farming relate to them (cf. NRC 2002: 244; Thompson 2003b). Judgments about risks (including about who should assume the "burden of proof") are integral to the disputes, but they should be located within this bigger picture.

This discussion of whether to endorse P_3, or not, illustrates a general point, which also applies to P_4, made by Richard Rudner fifty years ago (Rudner 1953; see also Douglas 2000; Lacey 2003b; section 3.4). Not all efforts to appraise theories (hypotheses) lead to acceptance (or rejection) of theories, or even to the judgment that sufficient evidence is not at hand to warrant acceptance but that further investigation is needed. Theory (hypothesis) appraisal, then, may eventuate in endorsing a theory (hypothesis). Endorsements are needed at the moment of application, although they are not simply judgments about the benefits of a proposed application (typically having to do with its side effects or alternatives to it) or the significance of a theory. I offer the following general statement:

> One (or a community) *endorses* a theory (hypothesis), T, when one judges that it manifests the cognitive values sufficiently highly so that responsibly applying it (or acting informed by it) does not have to take into account either that further research might lead to its rejection, or that, if T were false, applying it might lead to possible negatively valued consequences or threaten the manifestation of the values one holds.

Different communities may endorse conflicting propositions without violating any of the canons of empirical inquiry and while engaging in research that aims to test the degree of manifestation of the cognitive values in theories (hypotheses).[18] Clearly, as discussed, endorsements should be based on rigorous empirical inquiry. But the appropriate evidential standards to be utilized must reflect (social) value judgments. In situations where there is consensus or lack of consciousness about social values, or where a dominant party thinks its values are beyond contestation, the role of the values is likely to be overlooked, and endorsement is likely to be mistaken for sound acceptance. Then, illustrating one of the mechanisms of departure from impartiality and one that often mars discussions about the value of TGs, the different epistemic status attaching to judgments of efficacy and legitimacy may not be recognized. It seems to me to be an open question whether or not all matters that are now matters of endorsement (where conflict reigns) can in principle be transformed eventually—after exhaustive empirical inquiry—into a situation where one or other of the conflicting theories (or a successor to them) becomes soundly accepted.

NOTES

1. In this chapter, in order to contain the scope of the discussion, I address only environmental risks, leaving health risks aside. (There is not always a sharp separation of health and environmental risks. Certain alleged environmental risks would entail health risks also: e.g., bacteria becoming resistant to antibiotics engineered into plant genomes; environmental pollution creating health problems.) I have modified the statements of P_3 and C_3 accordingly.

Note also that, in order to avoid multiplying cumbersome endnotes, I have placed in the appendix (appendix: 2) several passages that are relevant to my discussion, but not woven integrally into the argument.

2. Accepting the need for risk assessments, however, leaves open questions about the adequacy of standard risk assessments, and even whether or not they have actually been conducted in some cases. Sometimes, no doubt, corporations, in search of quick or larger profits, will seek to cut the testing process short and to manipulate or bypass the regulatory system, or farmers, finding the regulatory system too burdensome, will ignore it. On the other hand, sometimes critics will exaggerate any suggestion of a risk (however tenuous) and mislead about the threat it poses, suggesting that adequate risk assessment is never carried out. These are failings that should be severely criticized, but the polemical context they may create should not be permitted to cloud the relevant issues about risks.

3. The strength of this presumption in favor of P_3 is dependent on the evidence in favor of Presupposition (b) (restated in the text). Those who question P_3 will presumably also question it. Note that evidence for (b) could not be obtained by research (like that conducted under materialist strategies) that dissociates from the causal nexus of undesirable side effects. One might also think that, rather than P_3 gaining backing from (b), independent evidence in favor of P_3 would be (together with evidence about the lack of relevant risks occasioned by other techno-scientific developments) part of the evidential base for endorsing (b).

4. This may be difficult to discern clearly when "scientifically grounded" solutions to problems are dissociated from the causal networks that create and sustain the problems. Often this has the effect that the socioeconomic conditions needed to implement a solution effectively are rendered invisible.

5. Item (i) is consistent with the fact that many possible hazards are known and have been subjected to risk assessment studies. It is intended to suggest that as yet unknown ones, which would be discovered (if they exist) if certain kinds of research were conducted, might be more problematic.

6. In actual fact, many on the P-side (including scientists) are driven by ideology (commitment to the modern valuation of control, or values of capital and the market), or self-interest, and they will be impatient with the carefully formulated claims being considered here. In the text I am identifying the P-side with the views of proponents who have attended explicitly and with nuance to questions of evidence (e.g., NRC 2002).

7. See appendix: 2, "Principle of Substantial Equivalence," for reference to a viewpoint about burden of proof that pertains to risk assessments concerning human health but not the environment.

8. Recent studies conducted in England suggest to some people that critics may have already successfully carried the burden of proof for some TG crops in some environments (see appendix: 2, "Recent Studies in the UK").

9. "One of the problems is that the companies [five conglomerates] have done nearly all the research on the crops' safety on their own or financed it elsewhere. If they want to build consumer confidence, they should embrace independent tests of the products' safety and impact" (*NYT* 2003b). These matters are exacerbated in the light of recent allegations of political interference (in the United States) in the decisions of regulatory bodies and in the dissemination of scientific results (*NYT* 2004a; Glanz 2004; McNeill 2004). (While none of the specific allegations have concerned TGs, they nevertheless raise questions about the extent of the interference and weaken trust in the declarations of these bodies.)

10. The C-side's argument is that what has been done to date does not amount to carrying the prior burden adequately. Given that, it will defer judgment on this counterclaim of

the P-side, until it sees how the risk assessments are improved. It might be more sympathetic to the counterclaim if the proponents of TGs were as ready to take or support action against those who violate regulations (e.g., the Brazilian farmers who have illegally planted smuggled TG seeds) as they have been against those who violate patent claims (Monsanto has many law suits against farmers who have grown their seeds in violation of their patent claims).

11. Compare: "The security of [TG] products has been proved scientifically, and no relevant argument has been raised disputing this. Transgenics are, in fact, more secure and efficient than conventional foods and also with respect to environmental impact" (Pavan 2003, my translation; Crodowaldo Pavan is a distinguished Brazilian geneticist).

12. For opinions about how well the P-side has carried the prior burden of proof, see appendix: 2.

13. There will be little interest in the middle ground where the modern valuation of control is strongly held, and also (typically in the same places) where the values of capital and the market are held. For these value-outlooks (via the presuppositions of the modern valuation of control) endorse the presumption in favor of endorsing P_3 usually so strongly that the burden of proof, put on the C-side, is beyond the limits of empirical investigation. This is effectively to accept P_3 on the ground that it is (or follows from) a presupposition of holding these values, and thus it is in violation of impartiality. There are also critics of TGs who effectively reject P_3 in violation of impartiality, for example, some of those whose opposition is founded in religious conceptions (appendix: 1)—they might verbally accept the middle ground, but then exploit it, constantly changing their interpretation by bringing up for consideration new "risks" every time the P-side deals adequately with a potential hazard.

14. In discussions, sometimes scientists (who are outspoken in support of the commercial release of TGs in Brazil) have responded to my arguments as follows: "I favor using TGs (they offer great benefits and per se occasion little unmanageable risk), but I oppose patenting of biological material." But the fact that most of the TGs in actual use are patented does not lead them to oppose using *these* TGs.

15. Similar concerns (as well as others) were evoked in Zambia in 2002 when, facing a potential food crisis, it refused to receive food aid that involved shipments of TG corn. "Zambia currently has no regulatory system or appropriate infrastructure to cope with the scientific assessments that should accompany the introduction of GE products. There is still great uncertainty about the safety of GE foods . . . for the environment. This led the Zambian government to evoke the Precautionary Principle. . . . Our environmental concerns were based on the fear that traditional corn varieties could be genetically contaminated. Since the aid had come in the form of whole grain, some recipients would likely save a portion of it for planting. This could lead to loss of agricultural diversity in Zambia" (Lewanika 2004: 87–88; see also Benbrook 2003).

16. I have portrayed the P-side as overshadowed by commitment to the modern valuation of control, which (in turn) is reinforced by holding values of capital and the market. Often its proponents not only endorse Presupposition (d) of the modern valuation of control (that there is no significant potential for value-outlooks that do not contain the modern valuation of control, but also the parallel one, that there are no significant possibilities outside of the trajectory of capital and the market (chapter 11). Then, the claim that relations of capital and the market (as distinct from matters open to legislative or regulatory change within the structures defined by these relations) are sources of risks seems odd; they frame the trajectory of the future: that is the way the world is! But investigating whether or not these relations are sources of risks is part of testing the new presupposition just stated.

17. I mentioned (section 6.3) allegations made by some on the C-side that the P-side engages in unethical conduct. One aspect of these allegations is that agribusiness corporations (and scientists who are sponsored by them) attempt to suppress evidence that there are serious risks (see Dalton and Diego 2002). It has been said, for example, that powerful agribusiness have organized pressures against journals that have published critical material about risks and the academic institutions that employ the researchers who have produced it. Both Losey (monarch butterfly case) and Chapela (Mexican "pollution" of maize landraces case) have complained of (what they consider) undue pressures exercised against them and their publications (Dalton 2004; Tokar 2004). I am not placed to evaluate the substance of these allegations. I mention them because most TGs are both objects for scientific inquiry and commodities; and interests linked with commodities can be in conflict with those of scientific research. This needs to be kept in mind in all discussions of risks and alternatives.

18. Thus, it is constitutive of endorsing a theory (hypothesis) that it be the outcome of "satisfactory performance of certain kinds of social interactions" (cf. references to Longino in section 1.2; Lacey 2005a), that it be produced under certain social relations (that embody specific social values) that have been cultivated among investigators. This is an important difference between endorse and soundly accept in accordance with impartiality. Note that sound acceptance is sufficient for endorsement (*SVF*: 71–74).

Chapter Ten

Alternative ("Better") Forms of Farming

P_4 There are no alternative kinds of farming that could be deployed instead of the proposed transgenic-oriented ways without occasioning unacceptable risks (e.g., not producing enough food to feed and nourish the world's growing population), and that reasonably could be expected to produce greater benefits concerning productivity, sustainability, and meeting human needs— "transgenics are necessary to feed the world."

C_4 Agroecological methods (and other alternatives) can be and are being developed that enable high productivity of essential crops (and occasion relatively less risk); and they promote sustainable agroecosystems, utilize and protect biodiversity, and contribute to the social emancipation of poor communities. Furthermore, there is good evidence that they are particularly well suited to ensure that rural populations in "developing'" countries are well fed and nourished, so that without their further development current patterns of hunger are likely to continue.

Arguments about the legitimacy of developing and using TGs must involve appraisals of benefits and risks. They must also draw upon appraisals of the productive potential and other potential benefits of alternative forms of agriculture. Risk-benefit analysis is necessarily comparative. Even if TG-oriented agriculture has all the benefits claimed by the P-side (chapter 8), including that it can produce food sufficient in quantity to feed the world's projected future population, there may still be "better" ways of doing so. Specifically, there may be alternative ways, or a multiplicity of complementary locally specific alternative ways, that simultaneously are productive of foodstuffs, environmentally sustainable, and strengthening of local communities, that are more in tune with the values and aspirations of communities of poor peoples and their variation with place and culture, and that are able to play an integral role in producing the food necessary to feed the world's growing population. Indeed, producing sufficient food to feed everyone does not ensure that everyone will be fed; massive hunger persists today, although enough food is being currently produced to feed everyone (section 5.5). Abolishing hunger is not simply a matter of increasing agricultural production, since hunger and how it is distributed is a consequence of the structure and functioning of agroecosystems (section 9.5); and there are reasons (potentially open to rebuttal!) to doubt (section 9.4; section 9.5) that it could it be the eventual outcome of agricultural research that treats seeds and their products simply as biological entities open to molecular biological investigation in dissociation from their ecological and social dimensions.

If agroecology (in combination with other approaches) provides a promising alternative, risk-benefit analysis that compares TG-oriented methods only with conventional high-intensity methods is inadequate. On the other hand, if there are no alternatives that can play a significant role in meeting the world's food needs and addressing the problem of hunger, then the anticipated benefits of developing and using TGs would legitimate taking greater risks for, it may be argued, whatever these risks may be, they pale in comparison with the risk that not enough food would be produced. That is why endorsing P_4 is so crucial to the argument of the P-side, and endorsing a version of C_4 to that of the C-side.

P_4 has the same logical form (negative existential) as P_3. Investigating it raises the same kinds of questions, as discussed in section 9.6, about the relevant strategies to adopt and the interplay of cognitive and social values in endorsing or not endorsing it. There are also issues about "burden of proof" (cf. section 9.4). The C-side will, and should, readily accept the (prior) burden in this case. When considering risks, there is considerable merit to staking out the middle ground, and so (while waiting for the results of specified research projects) to adopting the precautionary stance that takes seriously that there may be as yet unknown risks. To appeal to as yet unknown "better" agricultural methods, however, would be to deny legitimacy to uses of TGs simply on the basis of ungrounded hopes. The argument of the C-side collapses if concrete promising alternatives are not at hand, actually in use, favored by some movements of farmers, and open to improvement that is informed by systematic empirical inquiry. C_4 may be interpreted as affirming that the prior burden has been successfully carried, that the empirical record supports the promise of agroecological (and related) forms of farming.[1]

10.1 AGROECOLOGY: THE EMPIRICAL RECORD

There is ample evidence that agroecological farming practices can be effective along all four dimensions of "sustainability": productive capacity, ecological integrity, social health, and cultural identity (section 5.4). Miguel Altieri's book, *Agroecology: The Science of Sustainable Agriculture* (Altieri 1995), offers a veritable encyclopedia of successful agroecological projects. Additional examples can be found in de Grassi (2003); he compared TG crops grown in Africa (sweet potato, maize, and cotton) with crops grown following a variety of traditional approaches (deploying many typical agroecological methods) according to the criteria: "demand led, site specific, poverty focused, cost effective, and institutionally and environmentally sustainable," and (for the most part) the traditional approaches fared better. Furthermore, research conducted under agroecological (and related) strategies has been fruitful. Again Altieri's book documents this thoroughly. Often (see section 5.4), research conducted under agroecological strategies requires that farmers assume important roles.

Experiments in "Participatory Breeding"

This is well illustrated and confirmed in recent research on "participatory breeding" of crop plants, in which, for example varieties of maize have been developed—using traditional methods of selection, supplemented by "scientifically" developed and tested criteria for selection, and with the participation of both professional breeders and groups of small-holding farmers—that have high productivity under conditions of high environmental stress, for example in nitrogen-deficient soils (Machado and Fernandes 2001). Brazilian agricultural scientists Altair Machado and Manlio Fernandes report on the development of a variety of maize of this kind ("*Sol da Manhã*") that is now being marketed by the farmers who participated in its development, in the state of Rio de Janeiro. Another Brazilian agricultural scientist, Miguel Guerra, writes that this is not an isolated instance:

> An NGO (AS-PTA[2]) is working to recover and improve local varieties (landraces) with the method of "participatory breeding." In the west of the state of Santa Catarina some local varieties of maize yield more that 5 tons/hectare. This is higher productivity than that gained with the best hybrid. In several parts of Brazil we have similar experiences. It is very important for Brazil that, without using the high performance varieties developed in the green revolution that require stable climatic conditions and large amounts of inputs (fertilizers, pesticides, etc.), we can recover and improve local varieties that are well adapted to specific eco-regions. Being well adapted means being able to sustain good levels of yield even (and mainly) in the presence of biotic and abiotic stress. (Guerra, personal communication, March 2, 2004)

Alternatives to Monocultures

Some recent studies (of "ecological" approaches) are also interesting. Sometimes these are presented against the background of the accumulating ecological and social problems of high-intensity farming, and they suggest that research

214   *Chapter Ten*

limited to that conducted exclusively under materialist strategies is shortsighted. Consider the following comment from David Tilman:

> It is not clear which are greater—the successes of modern high-intensity agriculture, or its shortcomings. . . . The successes [e.g., of the green revolution] are immense. . . . But there has been a price to pay, and it includes contamination of groundwaters, release of greenhouse gases, loss of crop diversity and eutrophication of rivers, streams, lakes and coastal marine ecosystems. . . . It is unclear whether high-intensity agriculture can be sustained, because of the loss of soil fertility, the erosion of soil, the increased incidence of crop and livestock diseases, and the high energy and chemical inputs associated with it. The search is on for practices that can provide sustainable yields, preferably comparable to those of high-intensity agriculture but with fewer environmental costs, . . . that incorporate accumulated knowledge of ecological processes and feedbacks, disease dynamics, soil processes and microbial ecology. (Tilman 1998)[3]

He then cites a few experiments in which "ecological" alternatives, compared directly with conventional high-intensity methods over a ten-year period, produced comparable crop yields while simultaneously improving soil fertility and showing fewer detrimental environmental effects. In another paper, he introduces a set of studies that show that "greater diversity leads to greater productivity in plant communities, greater nutrient retention in ecosystems and greater ecosystem stability" (Tilman 2000; see also Tilman et al. 2002; Tilman and his colleagues are authors of numerous studies over the past decade, elaborating different dimensions of these issues, published in journals such as *Nature*, *Science*, and *Ecology*; Tilman 1998; 2000, both being introductions to sets of articles on similar themes, draw upon an extensive recent literature.)

Other experiments also show the promise of methods that do not use monocultures. Zhu and his colleagues, conducting research on rice crops in China, demonstrated that "a simple, ecological approach to disease control [intercropping of a few different varieties of rice] can be used effectively at large spatial scale to attain environmentally sound disease control" without loss of productivity compared to high-intensity farming of monocultures (Zhu et al. 2000). Martin S. Wolfe, an editorial writer of *Nature*, commenting on this result, after observing that long ago Darwin was aware that mixed cropping (of wheat) is more productive than monocultures, asked: "Why is the mixed approach not widely used?" He answered with the rhetorical question: "Is it just too simple, not making enough use of high technology?" (In my interpretive framework, is it ignoring P_1?) He continued: "Variety mixtures may not provide all the answers to the problems of controlling diseases and producing stable yields in modern agriculture. But their performance so far in experimental situations merits their wider uptake. More research is needed to find the best packages for different purposes and to breed varieties specifically for use in mixtures. . . . Mixtures of species provide another layer of crop diversity, with half-forgotten advantages waiting to be exploited in contemporary approaches" (Wolfe 2000).

Referring to his own and related studies, Tilman cautiously points out that they need to be backed up by further results in order to test the potential generality of the viability in practice of such alternatives. These studies, as well as the experimental and practical successes of participatory breeding, complement and (to some extent) confirm Richard Lewontin's theoretical argument (Lewontin and Berlan 1990) (and Shiva's criticisms of the green revolution—section 7.2) that developments of hybrids were not necessary to produce high-yielding varieties (of corn):

> The nature of the genes responsible for influencing corn yield is such that the alternative method of simple direct selection [that is in continuity with traditional farmer-selection methods] of high-yielding plants in each generation and the propagation of seed from those selected plants would work. By the method of selection, plant breeders could, in fact, produce varieties of corn that yield quite as much as modern hybrids. (Lewontin 1992)

Then, in order to emphasize that gaining increased yields need not have depended upon furthering the process of turning seeds into commodities,[4] he adds: "The problem is that no commercial plant breeder will undertake such investigation and development because there is no money in it." Apparently farming in which seeds are constituents of sustainable agroecosystems (and investigated as such) is not necessarily deficient in productivity.

Endorsing "There Are Better Alternatives"?

Does the empirical record just sampled add up to a case for endorsing C_4? The productive potential of agroecology is much contested, and there is a wide array of opinions about it.

At one extreme there are those, who endorse P_4 (and consequently P_3) without even discussing (or being aware of) the empirical record. They tend to be either in the grip of the modern valuation of control, and so (accepting P_1) do not regard the studies cited (since they are not limited to materialist strategies) as properly scientific, or to be impressed by the power and trajectory of contemporary market forces which admit little space for alternatives like agroecology. Then, (for them) the studies do not need to be appraised and can be ignored (see section 10.6 below). Insofar as they do not rebut alleged evidence in favor of C_4, however, their endorsement of P_4 does not have the support of empirical evidence. By identifying science with research conducted under materialist strategies, they put investigating P_4 outside the realm of science, so that endorsing P_4 functions effectively as a presupposition of what counts as scientific inquiry (in this area), rather than a result endorsed within it. (I have never seen a considered discussion by spokespersons of the P-side that attempts to rebut, or to reinterpret, the empirical record sketched above.)

Others acknowledge a role for agroecology, anticipating that TG methods will have to be supplemented by or fit in with other methods (ABC 2003; FAO 2004). For some of them, that role will be only a small one that requires little more than a footnote to P_4. These might concede that, pending adequate economic development in a region, agroecology is needed for some small producers (see discussion of possible reconciliation below in section 10.3); or they might see that its main potential is connected with "organic farming" that serves affluent minorities, and thus that it can serve a niche but not a mass market (Human Development Report 2001). Clearly empirical questions are involved here, and they should be addressed with well-planned investigation. Zhu's study speaks to both versions of the "small role" claim, though one study (it is the only one I know of that has been conducted on a large scale) cannot be decisive (Tilman 1999). One may wonder what the empirical record would be like if agroecological research had been (or were to become) supported with resources comparable to those made available for research and development of golden rice. As I also argued (section 9.4) in connection with P_3/C_3, the scarcity of evidence for C_4 (if that is how one interprets the available empirical record) counts as evidence against it, only if sufficient relevant research has been conducted. In this context, Tilman's words strike me as wise: "Research is needed that pursues all reasonable approaches to this problem. The apparent distrust between conventional and 'ecological' schools of agricultural thought must not blind either side to novel insights, nor slow the development of solutions to a global problem" (Tilman 1999). We just do not know what the extent of the potential of agroecology is, and also we do not know that the promise of TGs like golden rice can be fulfilled. We cannot know, in either case, unless further research is carried out. Nevertheless, the evidence supports that agroecology can meet the needs of significant numbers of small-scale farmers and their communities who tend to be bypassed in mainstream "development" projects and to be displaced following the growth of the practices endorsed by agribusiness; and the productive potential of agroecology is much greater than what has been realized to date.

More Research on "There Are Better Alternatives"

P_4, thus, lacks at the present time the empirical credentials needed to support its use in legitimating the universal value of TGs or questioning the potential or the value of agroecology. I'll leave it to those with practical experience in agronomy to judge whether or not the current empirical record adds up to a compelling case for endorsing C_4. If it does not, that might be either because there is compelling evidence that C_4 is false, or because insufficient research has been conducted. Without conducting further research under agroecological strategies, it cannot be settled which of the options holds. Only if such research is conducted, and the evidence gained fails to provide further support for C_4, would there be strong empirical grounds for endorsing P_4. So, even if there is not a compelling case to endorse C_4 at the present time, there is a compelling case to submit it to more intensive empirical scrutiny. That is enough to undercut the role currently played by P_4 (and consequently also P_2 and P_3) in arguments for the legitimation of widespread and intensive uses of TGs and for prioritizing research on TGs. On the other hand, since it recognizes uncertainties about the productive potential of agroecology and so cannot rule out some (even major) role for TGs in the agriculture of the future, it is not enough to deny value to conducting research and development on TGs or, following risk analysis carried out in the light of the precautionary principle, to using some varieties of TGs under special conditions (if they can be identified) where, for example, there is little promise of agroecological methods being effective.

It is true, of course, that conducting more research under agroecological strategies stands to serve interests informed by the values of popular participation,[5] and that, insofar as conducting such research requires farm lands that cannot be put to use by interests related to agribusiness, pose impediments to the expanded manifestation of the modern valuation of control and the values of the market. Nevertheless, the stipulated aim of science (section 3.2) and the interest of augmenting the manifestation of neutrality support the importance of conducting such research. In this case, the interests informed by the values of popular participation and the cognitive interests of scientific practices coincide.[6]

10.2 DEFENDING "TRANSGENICS ARE NECESSARY TO FEED THE WORLD"

The P-side will be wary of this argument. It might deny that the proponents of agroecology have adequately carried the prior burden of proof. Often the green revolution is its point of reference. The actual achievements of agroecology, it suggests, hardly compare with the enormous increases of food productivity that it enabled; and biotechnologically informed developments are hailed as its successors. From this perspective, the alleged potential (but not well enough demonstrated) productive improvements of agroecology—and arguments about what might have happened if, from the outset, research on agroecology had been comparably funded (Lewontin and Berlan 1990; Shiva 1991; section 7.2)—may seem to be simply irrelevant; then, it is easy to dismiss those who draw upon them as ideologically motivated (McGloughlin 1999; *AgBioView* 2004: numerous entries).

A variant of this response might be to modify P_4 by replacing "that reasonably could be expected to produce greater benefits" with "that have been demonstrated to produce greater benefits"; and then to argue that, given P_2 (and P_3), we do not need to explore the potential of alternatives and that it would be a misuse of funds to do so. Some important scientific (ABC 2003) and agricultural organizations (FAO 2004; Pinstrup-Anderson and Schioler 2000) have lamented that the potential for developing and using TGs for the sake of meeting the needs of poor farmers has been underexploited, and they recommend giving urgency and priority to developing this potential. Given the promised benefits of TG-oriented farming, why waste resources and spread our efforts thin trying to do better and, in doing so, risk impeding the realization of the benefits? The modified P_4, however, is not strong enough to sustain the argument for legitimation, for I argued (chapter 8 and chapter 9) that P_2 and P_3 are adequately defended only if P_4 is strongly endorsed empirically. Furthermore, keep in mind: first, it can be claimed today at most that there is potential for using TGs for meeting the needs of the poor, but an as yet untested potential. If the potential is not realized recourse will have to be had to the alternatives, so better that they not be left undeveloped. Second, there would be a significant time gap before the potential of using TGs for these ends could be put into practice (section 8.3)—meanwhile, the alternatives are available. Third, at least in some situations, there is evidence to back that the alternatives are more promising than the TG options (de Grassi 2003).

The P-side often charges that the questioning of P_4 (and its empirical credentials) is only ideologically motivated, and this charge dies hard. Why? In the first place, because questioning P_4 involves also questioning P_1. Since its proponents tend to identify science with research conducted under materialist strategies, they tend to interpret criticisms of the legitimation of a technoscientific advance as taking an "anti-science" (ideological) position (Borlaug 2000). Against this, without denying that the charge may sometimes be accurate, I have argued (chapter 7) that the aim of science should lead to endorsing C_1 rather than P_1 and that, when (in conformity with P_1) science is identified with research conducted under materialist strategies, its trajectory cannot be in the direction of higher manifestation of neutrality, because of the privileged place of its applications for value-outlooks containing the modern valuation of control. (It runs the risk of clashing with impartiality, too, when effectively C_4 is rejected because it cannot be investigated under materialist strategies.) It is not "anti-science" to endorse C_1 and to propose that P_4 be investigated empirically under strategies, not all of which will be reducible to materialist strategies. On the contrary, it is supported by long-acclaimed values of scientific practices and institutions. In the second place, commitment to the modern valuation of control, reinforced by contemporary global-market institutions and policies that highly embody these values, largely explain confidence in the possibilities of TGs to solve major problems of the poor. Paraphrasing Wolfe: "Agroecology is just too simple, not making enough use of high technology!" (It does not accept P_1!) Consider the two questions: "How can we genetically engineer plants that are herbicide resistant?" and "How can we engage in farming so that the sustainability of agroecosystems is enhanced?" On my account, both are scientific questions, open to systematic empirical inquiry, but only the first is closely and immediately linked with techno-science. Strong endorsements of P_1 and P_4 are explained (as already maintained in section 9.2), not by their accord with available evidence, but by reference to their links with presuppositions of the modern valuation of control.

Then, insofar as the scientific status of the matter is concerned, it seems that the charge of "ideological motivation" amounts to little more than annoyance with the C-side's contesting the modern valuation of control. The charge functions, one might suggest, (consciously or not) to draw attention away from the central role of commitment to the modern valuation of control in the argument of the P-side and from the alliance between powerful scientific institutions and large agribusiness corporations who are the primary beneficiaries of current developments of TGs.

10.3 THE CHALLENGE OF AGROECOLOGY

The challenge posed by agroecology (and other farming alternatives) to the rapid and widespread utilization of TGs has four components. First, it represents an alternative form of agriculture that is productive and sustainable, and that is relatively free from direct risks to human health and the environment. Second, agroecology has strong roots in contemporary practices that are linked with movements that embody values (e.g., those of sustainability, section 5.4, and of "popular participation") that conflict with the dominant moral vision of neoliberalism and the modern valuation of control (section 6.3; appendix: 1, sources, con). It is not an "abstract" alternative, a merely speculative possibility, but a spreading practice that is being informed by a growing body of knowledge gained in systematic empirical inquiry. It is socially rooted. It matters to the movements who embody the values of popular participation, so much so that, if its growth were undermined (as it is likely to be if policies favoring the use of TGs become entrenched), this would be a serious blow to the democratic aspirations of the members of these movements. Note that the second component without the first could only provide an "ideological" challenge; motivation to engage in agroecology may derive from the second, but any value (beyond narrowly local value) that agroecology has depends on the first component. Third, it suggests that perhaps the main risk posed by implementations of TGs is the destruction of alternative forms of farming which have the potential to feed and nourish the rural poor, or preventing them gaining the resources needed for their further development (section 9.5).[7]

Fourth, it challenges the notion of "science" or "modern scientific knowledge" that is present in P_1 (and that informs the other propositions of the P-side)—not from an "anti-science" perspective, but from one that denies that science should be restricted to inquiry conducted under materialist strategies. Agroecology presents itself as an alternative, an alternative well informed by scientific knowledge, knowledge derived under agroecological strategies (which make use of, where appropriate, knowledge gained under materialist strategies). This is investigation—not bound, like biotechnological research, by the strictures of materialist strategies—in which the biology and social science are inseparably intertwined, where the latest and most sophisticated technology are not always pertinent, and where the possibilities of "sustainability," and not just decontextualized possibilities, can hope to be identified. (While not denying that TGs are a product of sound scientific knowledge, it questions the claim of P_1 that this kind of scientific knowledge is generally exemplary.) It is only when science is conceived in such an expanded way, as a practice that can be conducted under a plurality of strategies of which materialist strategies are just one (albeit of importance and with some subordinate role in all strategies), that all the premises of the argument can themselves be subject to scientific investigation. In particular, P_4 cannot be adequately submitted to empirical testing unless inquiry under a variety of strategies, including agroecological ones, is conducted. And unless it gains support from that inquiry, it fails to play its key role in arguments for the legitimations of using TGs on the grand scale now.

The competition between agroecological and biotechnological strategies (insofar as they frame research on genetic engineering), rooted in contested social values, concerns the kind of scientific knowledge that should inform practical applications, and thus it also concerns research priorities. Where the values of the market and of the modern valuation of control are contested, there remains no objection in principle to engaging in research under competing strategies which, if fruitful, can be expected to inform practices that will further the social embodiment of the contesting values; and current fruitfulness suggests that the limits of agroecological strategies have not been reached. The practices that express the values of "sustainability" cut across the grain of the neoliberal project, and in these days of market triumphalism alternate possibilities are easily discounted. Nevertheless numerous groups of small-scale farmers throughout the impoverished regions of the world have made great improvements in their lives and communities through implementations of agroecology, which has become an essential part of their struggle to maintain and develop their cultural heritage as well as to meet their material needs (Altieri 1995; Rosset 1999; 2003; AS-PTA 2004).

Can Transgenics-Oriented Farming and Agroecology Coexist?

In the light of the four-point challenge just outlined, I urge the provision of space (and even, currently, priority) for the development of agroecological strategies. This does not imply either the wholesale replacement of research under materialist strategies, in part because research conducted under agroecological strategies regularly draws fruitfully upon knowledge gained under materialist strategies (section 5.4), or (see below) altogether abandoning research on TGs—although the critique made by the C-side of potential benefits (C_2), and certainly of the benefits of the currently most widely used TGs (herbicide resistant and toxic to targeted insects), remains in place. Now, I have mentioned (section 8.3; appendix: 1, sources, pro) that on the P-side there are differences in the values held among different groups that lead to some divergences concerning P_2. Groups proclaiming "humanitarian" goals, often including CGIAR-linked researchers, international development agencies, and publicly funded agricultural research institutions in many "developing" countries, often downplay the value of what are said to be the benefits of the currently widely used TGs on the ground that they are of no value to poor farmers in "developing" countries. Their endorsement of P_2 is qualified and tied to the promises that TGs will be developed that will have pertinence to these farmers and their conviction that some of these promises will be redeemed. Like the C-side, the "humanitarian" groups on the P-side (despite the modus vivendi they have established with agribusiness, section 8.3) tend to be critical of the priorities and projects of agribusiness corporations, and they articulate giving primacy to the value of responding to the urgent needs and problems of the poor. Do these points of coincidence provide a ground for some measure of coexistence, constructive dialogue, or shared projects between the two sides?

The values of "sustainability" (section 5.4) and those of popular participation (section 6.3) undergird the C-side, and so, in addition to prioritizing research and development on sustainable agroecosystems, it wishes (to the extent possible) to empower local rural communities, strengthen the agency of their members, to develop their traditional practices, and to enhance their cultural values. What strategies should be adopted in research, it asks having endorsed C_1, for the sake of furthering the manifestation of these values? In contrast, the "humanitarian" groups on the P-side—sometimes with a qualification or two that accepts that there is something important to be learned from traditional farming practices (McGloughlin 1999), other times clearly recognizing that TG approaches should be considered an indispensable part of a bigger package that will leave an important role for traditional approaches that are improved as a result of research (Serageldin 1999; Human Development Report 2001; FAO 2004)—for the most part endorse P_1. Instead of the question just posed, they ask: In what ways can the methods of biotechnology (including those of genetic engineering) be developed so that they can contribute to addressing the needs and problems of poor farmers in "developing" countries, for example, food production needs and chronic malnutrition in poor farming communities? They presuppose that science conducted under materialist strategies is a major part of the solution to problems facing poor communities (and they wish to be of help in generating solutions—they draw upon the ideal of neutrality here!); but, while they recognize the "reality" of the regime of IPR, they accept neither agribusiness dominance of research in biotechnology nor that the only access to seeds for farmers be through the market. Thus, the "humanitarian" groups conduct research aiming to develop TGs that can, for example, produce vitamin-enhanced rice, resist locally virulent diseases, or grow in soils that are water-depleted or high in salt content, and in this way aim to provide techno-scientific solutions to major problems of poor and marginalized farmers.

The C-side, in contrast, insists that such proposed techno-scientific solutions not be dissociated from the ecological and social contexts of their implementations, or from the historical causal nexus that created the problem to which the humanitarian groups wish to respond (Shiva 1991), and it offers alternative (agroecological and other) solutions that are likely to be available especially in tropical countries (such as Brazil) that are rich in biodiversity (Guerra et al. 1998a). I am not suggesting that at the present time the C-side can offer alternative solutions for all the problems that the "humanitarians" hope to address. In some situations it has already offered some, for example, participatory breeding experiments developed plants that have high productivity in nitrogen-deficient soils (section 10.1; see also de Grassi 2003). Whether or not, and under what conditions, it will be able to offer them can only be discovered in the course of research, and that depends on gaining the necessary socioeconomic conditions to carry out the research. The evidence surveyed above supports that the promise of agroecology extends well beyond its current actual achievements. Nevertheless, we do not yet know the limits of the productive capacity that could (under favorable conditions) be achieved by agroecological (and related) methods, and the potential suitability of these methods for all ecological conditions. There is considerable disagreement (among those who are familiar with agroecology) on these matters. I have talked, for example, with agricultural scientists in Brazil who think that, while agroecology is necessary for the small producer and therefore indispensable for feeding everyone, it is doubtful that agroecological methods will be developed to the point where they will generate the capacity to produce sufficient food to feed large concentrated urban populations, and to address adequately certain problems (e.g., of fungi in permanently humid climates) confronted in agricultural production (for other problems, see Gewin 2004). Given the disagreement, minimally there remains a case for continuing research on TGs; and, more strongly, there seems to be plenty of room for constructive dialogue (within a framework provided by C_1) and comparative empirical investigation here.[8] The unfolding of constructive dialogue, however, is inhibited by the fact that, under current conditions it cannot be expected that there will be any uses of TGs that will be free from constraints imposed by the regime of IPR. (I mentioned, in section 6.1, that the C-side does not stand opposed to agricultural biotechnology in general. This is a basis for constructive dialogue between agroecologists and those with formation in biotechnology, but I will not develop this point here.)

Constructive dialogue requires full recognition of the agroecosystems in which farming is practiced and, thus, at least the provisional endorsement of C_1. Endorsing C_1 now is not incompatible with, later, after examining the research record—should it show that research conducted under strategies that are not reducible to materialist strategies have not been, or no longer promise to be fruitful—coming to endorse P_1. Remember the questions posed respectively by the C-side and the "humanitarians" on the P-side: "What strategies should be adopted in research in order to further the manifestation of the values of 'sustainability' and popular participation?" "In what ways can the methods of biotechnology (including those of genetic engineering) be developed so that they can contribute to addressing the needs and problems of poor farmers in 'developing' countries, for example, food production needs and chronic malnutrition in poor farming communities?" Posing the former question neither presupposes, nor precludes, that its answers will be informed by answers to the latter. On the other hand, when it is presumed that there must be positive answers to the latter question, issues of sustainable agroecosystems become subordinated to the search for techno-scientific solutions to problems; and then there can be no basis for constructive dialogue.[9]

Conditions for the Empirical Appraisal of P_4

Whatever the possibilities of coexistence between agroecology and some uses of TGs may be, credible investigation of P_4 must address (and attempt to rebut) the empirical record cited by the C-side against it, or (more likely) argue that the prospects for agroecology are inherently severely limited. Such supposed limits cannot be identified unless more research is conducted under agroecological strategies and appropriate resources are provided for this research. Research on the possibilities of sustainable agroecosystems requires the cultivation of agroecosystems, for the object of investigation cannot be reproduced on a small or simplified scale in the laboratory. The productive potential of agroecology can be empirically appraised in practice only if actual agroecological farming practices are intensified and expanded. In order to test P_4, therefore, resources need to be provided to support the intensification and expansion of these practices; but providing them runs into conflict with the thrust of the global economy itself, whose logic favors the rapid and immediate introduction of TG-intensive farming. This thrust serves to undermine the conditions (the availability of productive and sustainable agroecosystems) needed for the scientific investigation of a presupposition (P_4) of its own legitimacy. If it remains unchecked, it may obliterate alternative forms of agriculture and the movements that want to develop them; and then (as pointed out in section 9.5) P_4 might become true because a project, whose legitimacy presupposed it, actually undermined the grounds for alternative approaches. That would be a great moral tragedy, and it represents the greatest risk of the introduction of transgenics (one not conceptualized in standard risk assessment studies).

There is also a sort of paradox threatening here: techno-scientific innovation undermining the conditions needed to investigate scientifically a presupposition of its own legitimacy. It comes from presupposing P_1, rather than coming to endorse it after having investigated the claims of propositions like C_4 and failing to gain evidence to support them. This threat of paradox may be relieved by my suggestion (section 10.6 below) that it is not so much scientific results, as socioeconomic commitments, that underlie the best arguments for the legitimations of using TGs intensively now. Nevertheless, any authority that science may properly exercise derives from the results of systematic empirical investigation. At the present time, that authority does not stand behind P_4. Perhaps the appeal to science made by the P-side arises out of the mutually reinforcing relations between holding the modern valuation of control (and thus giving privilege to research conducted under materialist strategies) and holding the currently predominant values of capital and the market (section 10.6), or it masks the lack of a generally convincing moral foundation for globalization, an effort to unnerve its critics, or a boundless faith in the powers of materialist strategies. Whatever it may be, the C-side, when it cites the promise of agroecology, is not running against established science. On the contrary, the intensification and expansion of agroecology are necessary so that there can be scientific research that investigates the possibilities for feeding everyone in the immediate and foreseeable future, and thus put us in a position where P_4 can be appraised with appropriate empirical rigor.

10.4 HOW TO CONDUCT SCIENTIFIC RESEARCH?

Throughout part I of this book I argued that there may be a mutually reinforcing relationship between adopting a particular kind of strategy and holding a particular value-outlook, that such relationships may contribute both to explain and to provide reasons for the adoption (or the predominance) of certain strategies, and that, consequently, it is important for the worldwide scientific institutions to provide conditions for research to be conducted under a plurality of strategies. (Moreover, I have repeatedly suggested, that determination of the range and priorities of this plurality should be a matter for democratic deliberation.[10]) Summing this up, I wrote at the end of chapter 2:

There are rich dialectical interactions among the questions: "How to conduct scientific research?" "How to structure society?" and "How to further human well-being?" Science may be appraised, not only for the cognitive value of its theoretical products, but also (without threatening this) for its contribution to social justice and human well-being.

This general analysis has been illustrated (and confirmed) by the analysis of the controversy about TGs offered here in part II, especially in the discussions of risks (chapter 9) and alternative forms of farming (this chapter). It has also shown that the democratic aspirations of movements that endorse the C-side are at stake, and that lends urgency to the questions: Under what strategies, what plurality of strategies, shall we conduct scientific investigation for the sake of appraising the legitimations of using TG crops now and of policies that make them central to the agriculture of the future? Under what plurality of strategies can we hope to move toward making judgments that accord with impartiality with respect to key propositions deployed in arguments for legitimations?

I have repeatedly emphasized that it is vital to distinguish between the strategies under which research is conducted and the theories (or knowledge proposals) developed under them. Choice of strategies identifies the objects of knowledge and the kinds of possibilities that one is interested in investigating; that choice tells us nothing about what actually is the case or is possible. Accepted theories identify the class of genuine possibilities in the domain of investigation; the criteria of cognitive appraisal have (or ought to have) nothing to do with the values that render the possibilities being investigated interesting (for application or other social role); they involve only relations among theories and relevant empirical data; established knowledge claims are not properly subordinated to social values or metaphysical or religious outlooks (impartiality, section 1.2). This holds for research conducted under agroecological strategies as well as under biotechnological strategies. Under the former we aim to identify the possibilities of "sustainable" agroecosystems (section 5.4), under the latter, for exapmle, the possibilities available for the genetic modification of crop seeds. These classes of possibilities cannot be investigated together under the same strategies. Research that is conducted virtually exclusively under materialist strategies will never address the former.

I think that it is sound science policy in a democratic society to institutionalize scientific research so as to permit, in fields (like agricultural science) where it is appropriate, a plurality of approaches to develop, with full awareness of how an approach may be linked with particular values—so that (making concrete general remarks made in section 2.1) (a) values will not play a covert role in accepting and rejecting theories or knowledge claims (like P_4), (b) value disputes become part of discourse within the worldwide communities of investigators—and scientists become free to adopt an approach because it can hope to identify possibilities that will serve the interests shaped by the values of "sustainability" and popular participation, and (c) science not consider itself as free from a measure of democratic oversight. The furtherance of the ideal of neutrality depends on this kind of institutionalization. Repeatedly we see deep links between the scientific ideal of neutrality and democratic ideals.

Where scientific research has become caught up in the quest for gaining IPR I fear that this will not be considered a serious proposal. In that context the claim for scientific respectability of research conducted under agroecological strategies, and making available for its conduct the necessary material and social conditions, will continue to evoke political controversy and often to be dismissed without a serious hearing. This is to the detriment both of science (and its alleged neutrality) and to the quest for social justice of numerous movements of small-scale farmers and their allies. The value of research conducted under agroecological strategies (and the questioning of the universal value of TGs) derives neither from nostalgia for a "more natural" way of life nor from "anti-science" sentiments. Rather, it derives both from solidarity with poor people whose movements are struggling to recover and enhance their personal and communal agency and from commitment to furthering the neutrality of scientific practices and the impartiality of claims endorsed with the authority of science. My argument clashes with the predominant self-image of the contemporary professional scientific community, which (endorsing P_1) tends both to give a monopoly to techno-science and (dissociating from the social/economic/cultural relations of research/development/application) to consider its products neutral and available to be drawn upon by interests connected with practically any value-outlook.

Broadly speaking (here I repeat and expand remarks made in the Introduction) there are two steps to my argument (hence the two parts of the book). The first, making the case for engaging in research under a plurality of strategies, is an argument in the philosophy of science, and it draws principally upon my statement of the aim of science (section 3.2) and explores how to further the manifestation of the widely acclaimed scientific ideals of impartiality and neutrality. In order to show that this plurality represents more than an abstract possibility, the argument needs to be supplemented by instances of fruitful (or potentially fruitful) research conducted under strategies that are not reducible to materialist strategies and that have mutually reinforcing relationships with values that contest the modern valuation of control. I used the case of agroecological strategies (in section 5.4, section 5.5) for this purpose. The strength of the argument in part I does not depend on holding any particular ethical/social values (apart from those implicit in the aim of science). That I have highlighted agroecological strategies does reflect my own commitment to the values of popular participation, but that commitment is irrelevant to the appraisal of the fruitfulness of these strategies (as are the facts that they are not reducible to materialist strategies and that these values contest the modern valuation of control). It may well be that holding certain kinds of values is a socially necessary condition for a researcher to recognize the interest of alternative strategies; if so, that would be the basis for an argument that different value-outlooks should be well represented in the institutions of science (Longino 2002).

The second step involves arguing for the crucial importance now of engaging in more research conducted under agroecological strategies. Here the ethical stakes at play in the TG controversies are important—both with respect to empirically appraising presuppositions of the legitimacy of using TGs intensively now, and to provide knowledge that can inform agroecological practices, which are of great interest to movements that hold the values of popular participation and which are being used ever more widely by their members. These values motivate engagement in research conducted under agroecological strategies (and motivate pushing for higher empirical standards to be met for endorsing P_3 and P_4) and, at the same time, engaging in that research furthers the interests of impartiality and neutrality.

The two-step argument, although it is based on a viewpoint in the philosophy of science, is not an abstract one, and it does not involve merely negative criticism of mainstream science, since it is also rooted in critical reflection on the practices of agroecology. It makes a positive case for the scientific significance of the knowledge that informs these practices and for the value of research that builds upon it. It is part of a philosophical "theory" that interprets and supports both the practices and the research conducted to inform them (as having a proper place—alongside others—within scientific practices), and defends their credentials from criticisms that they are "unscientific." So it is an argument that confronts the predominant self-image of contemporary science with the sound claims of an alternative practice. The strength of the argument goes hand in hand with the value and viability of the alternative practices.

10.5 ETHICAL DISCOURSE AND SCIENTIFIC INQUIRY

Where the modern valuation of control is held, there is likely to be a presumption made in favor of P_4, but that presumption cannot properly trump the claim that the C-side has carried the "prior" burden of proof, as (I suggest) is confirmed by the empirical record sketched in section 10.1. Unless there is a decisive rebuttal of this record (and, as I said above, I have never seen it seriously discussed by the P-side), the argument for the legitimations of using TGs on a large scale now will depend on empirical evidence being provided for P_4, and that can only be provided with further empirical scrutiny of the claims of C_4 and consequent failure to vindicate their promise. In the absence of such empirical scrutiny, I expect ethical discussion of TGs to continue to degenerate into the posturing that so often mars the controversy, with the P-side appealing to P_4 effectively because it "legitimates" their projects and does not challenge their power, and the opponents denying it in a show of ethical "bravado." Ethical discourse on this topic (as well as many other topics), for its authenticity, must draw upon the results of scientific inquiry, and, where the relevant results are not at hand, it must urge that the research be conducted.

The force of this point can be obscured by acceptance of P_1, however, since it effectively identifies "science" with "research conducted under materialist strategies, and the knowledge gained in that research." This makes the terms of ethical discourse even more difficult. Once one accepts that science is not limited to inquiry conducted under materialist strategies, it becomes apparent that, when science is restricted to the play of these strategies, it is not because of the requirements of science (systematic empirical inquiry) per se, but because science has been inherently linked with the modern valuation of control. Values are there, at the outset, in research on TGs (section 5.4): the values do not determine what are the possibilities open to genetic engineering, but they do account for the overwhelming emphasis on exploring systematically these kinds of possibilities. That is why the "humanitarians" on the P-side are mistaken when they propose that the key issue is how to shift the emphases of research and development (for the sake of application) of TGs from corporate interests to those of poor farmers (section 10.3). It is not sufficient to limit the ethical appraisal of scientific research only to its applications; it must also address strategic choices and priorities. The biotechnological research that has produced TGs is, from the outset, linked with the modern valuation of control (and consequently today with the values of the market-oriented global economy). Within the framework provided by this value-outlook, the P-side (including agribusiness corporations) is generally prepared to consider seriously ethical questions, especially those connected with risks to health and environment.

For the C-side, however, the matter of contention derives from the P-side holding the modern valuation of control itself, and its (and its presuppositions) providing legitimacy to create obstacles for alternative forms of agriculture. Both the biotechnological research that has produced TGs and agroecology provide scientific knowledge—knowledge that, respectively, informs different projects linked with different value-outlooks, among which there are vast inequalities of power. Ethical discourse should not abstract from the value-outlooks that lie behind it and that may prevent serious engagement with the issues. Nevertheless, at the same time, the legitimating presuppositions of value-outlooks (like those of the modern valuation of control—section 11.4) should be submitted to empirical inquiry to the fullest extent possible. The value-outlook, linked with the P-side, underlies the claim that TGs will contribute to meeting the world's food needs and to serving human well-being generally, and (P_4) that there are no other ways to do this. But, the C-side counters, the institutions that produce TGs have a major role in socioeconomic structures within which vast numbers of people are not fed, although enough food is produced globally so that there is enough food currently available to feed everyone (section 8.3), and, in continuity with this, the expansion of the use of TGs can be expected to drive more small farmers from their lands, and thus to exacerbate problems of hunger and social dislocation. Of course, if P_4 is true, there may be no possibilities for feeding the world without the role of TGs (and of the institutions that have produced them); and then the focus would have to turn seriously to prioritize the question of distribution. Thus, the authenticity of maintaining value-outlooks that contain the modern valuation of control requires investigation of P_4, and thus (at least temporarily) granting space for development of agroecology.

Self-interest may lead one (or a corporation or a government) to ignore this, but it remains a condition of authenticity—even more so, since the value-outlook of the P-side emphasizes the rationality of acting in accordance with the best results of properly conducted scientific inquiry. The moral presuppositions of the value-outlook linked with agroecology should be held to similar standards. The value of solidarity with popular movements of poor people, or affirming the values of popular participation, does not remove the necessity for empirical scrutiny of the possibilities of agroecological production. Certainly not enough is known today to maintain with confidence that TGs will not be necessary, so that empirical investigation of the risks of TGs remains important. But, as already pointed out, the significance of risks is less or greater depending on whether there are viable alternative forms of agriculture.

Agroecology shows great promise. Exploring that promise should be a high priority, but (section 8.3) agroecology does not involve a "blueprint" which can be proposed for implementation widely throughout the world. Rather it consists of a cluster of locally variant developments, whose potential can be tested locale by locale, and it occasions none of the risks of rapid large-scale transformations of farming practices. But, there are other kinds of "risks" for those who share the interests of agribusiness and those who hold the modern valuation of control, whose hopes (backed by large investments) are tied to the progress of science conducted under materialist strategies. If agroecology were widely implemented, following numerous and various models of successful practice, then the investments made in developing TGs might not be paid off and the ambitions of agribusiness and the powers of large landowners might be curtailed. The latter "risks" help to explain why movements committed to the values of popular participation, like MST in Brazil, may even have to face violence, both from the state and the vigilantes organized by landed oligarchs (Stédile 1999; Branford and Rocha 2002; Wright and Wolford 2003). They also explain why it is not surprising that there is little enthusiasm for agroecology in mainstream discussions. That, of course, is not evidence in favor of P_4, but it does point to the high stakes that are involved, and it brings into clear light the essential role of ethical/social value judgments in appraisals of benefits and risks. In any case, the promise of agroecology cannot be explored without the development of rural movements committed to the values like those of popular participation, and supporting those already in existence. It follows that putting resources into supporting rural movements that are committed to popular participation is required (and, in Brazil, this implies support for programs of agrarian reform) not only for attending to the needs of the rural poor now, but also in order to investigate P_4 empirically, and thus to be able to make more decisive ethical appraisals of TG-oriented agriculture.

Meanwhile, it is ethically legitimate to proceed with research and development of TGs, subject to careful, wide-ranging, and theoretically well-informed studies of risks, and to introduce the practical use of TGs step by step—after adequate risk assessment has been conducted, and subject to careful ongoing monitoring; and recognizing that adequate risk assessment needs to take into account important differences in the environments where TGs may be used. Agribusiness companies, however, have no interest in the small-scale introduction of TGs. Even so, large-scale TG-oriented agriculture cannot be ethically legitimated, on grounds that can properly claim the support of science, as a "global" policy before P_4 is thoroughly empirically tested—and that can be done only if non-global-market-driven modes of production, like those of agroecology, gain the opportunity to be developed more fully. Unless this is recognized, the prospects of strengthening democracy in many of the world's impoverished countries will remain remote.

10.6 GROUND OF LEGITIMATING CURRENT USES OF TRANSGENICS: SCIENCE OR MARKET INTERESTS?

I wrote above (section 10.3) of the sort of paradox that threatens the argument of the P-side: techno-scientific innovation undermining the conditions needed to investigate scientifically a presupposition of its own legitimacy. There is a way to relieve this threat, but that turns the P-side argument more into a sociopolitical one than one based in science. Recall Presupposition (d) of the modern valuation of control: "There are no significant possibilities that value-outlooks which contest the modern valuation of control will be actualized in the immediate future" (section 1.1). This may underlie the proposal that, somehow, it is misguided to urge the adoption of research strategies that bear mutually reinforcing relations with values that contest the modern valuation of control—misguided sociopolitically, not misguided from the perspective of the aim of science. This would not necessarily imply that science should give up commitment to impartiality, but it does imply subordinating neutrality to the prevailing social "realities," and autonomy (with respect to the general orientation of research) to the interests of the principal funding sources. The proposal begins to make sense of the fact that often, when (in discussions) I present the argument that P_4 lacks the support of empirical evidence, proponents of TGs react as if I am missing the point. Rarely do they challenge the evidence that has been accumulated in favor of C_4 (which I have acknowledged to need strengthening); they wonder why I am bothering to offer it at all.

Despite the rhetoric of science accompanying the P-side, I am inclined to think that its argument is driven, not so much by P_4, as by:

P_4a: There are no alternative kinds of farming—*within the trajectory of the socioeconomic system based on capital and the market*—that could be deployed instead of the proposed TG-oriented ways, without occasioning unacceptable risks (e.g., not producing enough food to feed and nourish the world's growing population), and that reasonably could be expected to produce greater benefits concerning productivity, sustainability, and meeting human needs; *and outside of this trajectory there are no genuinely realizable possibilities*.

An interesting and sophisticated version of this proposal can be found in some of Paul Thompson's publications. He argues that there should be a presumption in favor of "food biotechnology." In context, this implies endorsing a presumption in favor of a modified version of P_3 (like P_3a in section 9.3); he thinks that the P-side has adequately carried its "prior burden" in risk assessments of TGs, and that current risk assessment and monitoring practices (in the United States) are fundamentally sound. More importantly, his argument draws upon "the late twentieth century social bias in favor of technology . . . [the fact that] industrial societies are organized in ways that institutionalize the bias favoring technology [including TG technology]," and thus that any alternative approach will be extremely costly (Thompson 1997: 23–25; see also Thompson, 2003a; 2003b). I interpret this as endorsing the presumption in favor of P_4a in the contexts of the advanced industrial countries—so that (for him) the consequent legitimacy of widespread uses of TGs is relativized to current socioeconomic conditions in the advanced industrial countries.

According to this version of the argument, P_4a is presumed on the basis not of empirical evidence about the possibilities of alternative forms of farming, but of an assessment of socioeconomic trends that allegedly do not permit alternative forms of farming to gain significant space to flourish. It makes judgments about the possibilities of alternative forms by taking into account the social context of contemporary farming (it's just "realism"!). Now, the "bias" that Thompson cites is clearly connected to some of the presuppositions of the modern valuation of control; and P_4a may be interpreted as being grounded in Presupposition (d). The socioeconomic trends that permit little space to alternative forms of farming embody highly this valuation (as well as values connected with property and the market which shape the institutions that are its primary bearers today), and it is because they embody the modern valuation of control that they allow only that little space. Put like this, the presumption in favor of P_4a is derived from an empirically based social analysis and need not involve its proponents actually adopting the stated values.[11] Such analysis of actual trends cannot rule out, however, that there are possibilities (that might be actualized by movements which embody competing values) to counter those trends (chapter 11). Certainly, there are movements that question the inevitability of the actual trends and the values of the institutions which are restructuring the world under the "bias favoring technology," for example, from the perspective of the values of popular participation. They do not deny the actuality of the social trends just referred to and the powerful forces driving them, so that they recognize that no alternatives could be implemented without a struggle (section 11.3) and that the possibilities for alternatives (e.g., alternative forms of agriculture) are matters open to empirical inquiry. Not to take such movements into account is effectively to hold the stated values—perhaps regretfully and not wholeheartedly, yet—nevertheless—holding them. When movements that embody competing values are organized in "developing" countries and gain some political strength, and the presumption of P_4a (or Presupposition (d)) is posed against their endeavors, holding the values of capital and the market (and the modern valuation of control) is the key support of the presumption.

Rhetorically, P_4a is very powerful. It tends to be put forward—usually ignoring Thompson's relativization of it to the advanced industrial countries—as a "realistic" proposal, and it characterizes alternatives as outside of the trajectory of the present-day world, with no possibility of having roles in the society of the future, and so worthy of neglect, since (it is said) they are motivated only by a futile and irresponsible resistance to "progress," whose dynamic (at the present historical moment) reflects the movements of capital and the interests of the market and which permits (or tolerates) no resistance. This contributes to explain why it is so difficult to engage with the P-side regarding empirical investigation of P_4 and P_3 (C_4 and C_3), when socioeconomic relations are hypothesized as causal sources of risks and the underdevelopment of agricultural alternatives. (What is a source of risk for the C-side is the way things are for the P-side.) It is difficult to dialogue about P_4a; it represents a fundamental part of the self-understanding of the neoliberal project, and it is sustained (in part) not only by the deep embodiment of neoliberal values in contemporary institutions, but also by the use of its economic and political power in an ever-extending effort to control the world's socioeconomic spaces.

The role of P_4a contributes, also, to explain why many scientists who hold "humanitarian" values and are critical of most current uses of TGs (seeing them as principally serving the interests of agribusiness), nevertheless focus on the question: "How can we develop and utilize TGs in order to serve the interests of poor farmers?" and not on the more general question: "How can we develop agricultural research, and under what variety of strategies, in order better to

Alternative ("Better") Forms of Farming   233

serve the interests of poor farmers and their communities?" They presuppose the importance of TGs at the outset and not as a conclusion of research that is open (in principle) to a wider range of responses. Despite this, P_4a is open to empirical investigation, too, and issues about this will be taken up in the next chapter. Note, however, that the refutation of the final clause of P_4a would depend (in part) on the possibility of creating spaces for the development of alternatives within the currently market-dominated system, and that this could not happen without investigating empirically possibilities that are supposedly not realizable within the predominant trajectory, but which could (in principle (if there are any such possibilities) be investigated by research conducted under strategies (including agroecological ones) adopted so that the second question stated above could be addressed.

Variant Formulations

In Brazilian discussions of TGs, one hears a number of variations of P_4a, all of which depend on accepting it for their plausibility (Araujo 2001). (These variations have their obvious counterparts in discussions in other parts of the world.) One variation insists that the indispensable role of TGs in agricultural practices is already a fait accompli, so much so that there is no possibility of a change in direction; that expectancies about alternatives reflect only nostalgia for the past ways. "There is no return to the past," they say, as if the advocates of C_4 proposed doing so. We may put this:

P_4b There are no alternative kinds of farming that could be deployed instead of the proposed TG-oriented ways, without occasioning unacceptable risks (e.g., not producing enough food to feed and nourish the world's growing population), and that reasonably could be expected to produce greater benefits concerning productivity, sustainability, and meeting human needs—since it is a fait accompli, with no possibility of reversal, that TGs have become an integral part of leading agricultural practices; *the alleged promise of proposed alternatives reflects only an unrealistic desire to return to the past (or exaggerated extrapolation from small-scale forms of farming).*

A second variation emphasizes the important role of agricultural exports for the Brazilian economy:

P_4c There are no alternative kinds of farming—*that do not occasion such unacceptable risks as not being able to maintain and augment the productivity of crops that are essential to the Brazilian export economy*—that could be deployed instead of the proposed TG-oriented ways, without occasioning unacceptable risks (e.g., not producing enough food to feed and nourish the world's growing population), and that reasonably could be expected to produce greater benefits concerning productivity, sustainability, and meeting human needs.

I have wondered about what is the evidence that supposedly supports P_4c, since the productivity of non-TG soybean in Brazil has become greater than that of soybean crops (the larger part of which is TG) in the United States (Araujo 2003), its agricultural exports continue to grow rapidly, and the large EU market gives preference to non-TG products. Perhaps there is a sense, which might be based on accepting P_4a, that it is only a question of time until EU preferences will change and the market come to discriminate against non-TG products. Such a sense might also be based on accepting P_2; from newspaper reports it seems that the large numbers of farmers in the state of Rio Grande do Sul, who are growing crops of TG soybean from seeds smuggled from Argentina, expect to gain larger profits. There may also be a sense that it is inevitable that farming practices of the future will be informed by the latest developments of techno-science. Thence, paralleling P_4a, and drawing on the role that techno-scientific innovations play in strengthening the system of capital and the market:

P_4d There are no alternative kinds of farming—*informed by recent (and future) developments of techno-science*—that could be deployed instead of the proposed TG-oriented ways, without occasioning unacceptable risks (e.g., not producing enough food to feed and nourish the world's growing population), and that reasonably could be expected to produce greater benefits concerning productivity, sustainability, and meeting human needs; *and only farming practices informed by techno-science will remain viable (outside of small special niches) in the future.*

P_4d is reinforced by yet another variation, which holds that liberating TGs for farmers to use is necessary for the sake of strengthening the progress of Brazilian national scientific research.

P_4e There are no alternative kinds of farming—*that do not occasion such unacceptable risks as creating barriers to the development of science (of research and development of techno-science) in the nation*—that could be deployed instead of the proposed TG-oriented ways, without occasioning unacceptable risks (e.g., not producing enough food to feed and nourish the world's growing population), and that reasonably could be expected to produce greater benefits concerning productivity, sustainability, and meeting human needs.

Leading scientific institutions and bodies that support them (federal and state government supported foundations; the Ministry of Science and Technology), it is often said, have made the commitment to prioritize biotechnological research (in both agriculture and medicine) for two related reasons: (i) In this area of research Brazil could be competitive internationally; with priority support for it, the highest levels of Brazilian research can be strengthened. (ii) Biotechnological research will lead to discoveries that can be patented, and the national holding of patents will lead to lowering of the prices of agricultural products in the country; (more importantly) it will permit the development of TG technology that is especially appropriate for Brazilian conditions and diminish dependence on foreign institutions—and so contribute positively to the Brazilian economy. This can come about only if there is a close connection between scientific research and technological application. Liberating TGs for immediate use by farmers serves these two interests—scientific and economic—that are shaping current forms of development in Brazil. In this context, it is "natural" to interpret C_4, as well as any other expressions of opposition to immediate liberation of TGs for use, as against the development of Brazilian science. But that could be defended only on the assumption either of P_2 or P_4a. When the argument is recast in this way, we see that the charge that the C-side is "anti-science" comes to mean, not that it runs counter to the authority of well-established scientific judgments (expressive of impartiality and neutrality), but that it clashes with interests of leading professional scientists.

Replacing P_4 by P_4a is an interesting move, and it captures certain tendencies in the controversies about TGs. It also changes the terms of the discussion. Once the replacement is made, the authority of "modern science" can no longer be considered decisive. The strength of arguments for the legitimations of using TGs would now depend on whether or not there were good empirical grounds for endorsing P_4a—a matter that cannot be addressed in research conducted exclusively under materialist strategies.[12] In part III, I will explore what kinds of strategies need to be adopted in social science research in order to investigate empirically propositions like P_4a and presuppositions of the modern valuation of control.

NOTES

1. "Agroecology" tends to be used to encompass a variety of farming methods. It is not directly pertinent to the present argument to discuss whether or not it incorporates all the approaches that present themselves as alternatives to both TG-oriented and conventional high-intensity farming. An editorial of the Brazilian journal, *Agroecologia Hoje* (2, no. 10, August–September 2001) distinguished "organic," "ecological," "natural," and "biodynamic" approaches, and proposed grouping them together under the label, "bioagriculture." I do not suggest that agroecology is the only approach appropriate in the light of the values of popular participation. In this chapter I will use "agroecology" as shorthand for "agroecology and the variety of related approaches." See also Guerra et al. (1998a).

2. AS-PTA: Assessoria e Serviços a Projetos em Agricultura Alternativa, based in Rio de Janeiro. Its website (see bibliography) provides references (in books and journals) to numerous studies in agroecology. It is a useful supplement to Altieri's book.

3. Tilman is pointing to harm that has been caused by high-intensity farming, presumably harm that was not taken into account in earlier risk assessments and so not circumvented by regulations for its use. Some proponents of TGs maintain that TG-oriented farming can avoid, and even redress, these kinds of harm—a techno-scientific solution for harm caused by techno-scientific informed practices. (See Presupposition (b) of the modern valuation of control—section 1.1.) That is not Tilman's proposal.

4. Note that there is no evidence to date that TG's may be developed that produce higher yields than high yielding hybrids. Weiner (1990) offers a theoretical argument that they will not be.

5. That kind of research is not guaranteed to serve these interests. It may turn out that the research would quickly expose serious limits to the productive potential of agroecology. In that case P_4 would gain empirical support. No doubt, those who hold the values of popular participation would expect high standards of testing to be deployed before that would concur with this. But their endorsement of C_4 is subject to empirical constraint. This fits with the interplay of cognitive and social values in the context of endorsing hypotheses, but it does not subordinate cognitive to social values.

6. In contrast, neutrality is not furthered (and neither is impartiality) when a presumption in favor of P_4 rests effectively on holding the modern valuation of control, and when this leads to curtailing research being conducted under a plurality of strategies; section 1.3; section 5.5.

7. I have never seen this point addressed by the P-side. When I raise it in discussion, the response is usually dismissive and reflective of internalizing a kind of certitude with respect to propositions like P_4a (introduced in section 10.6) and Presupposition (d) of the modern valuation of control. (It also reflects, I think, various aspects of the violence—against people's knowledge and against poor people—diagnosed by Shiva, section 7.2, as consequences of conceding a monopoly to reductionist science.) What explains the certitude displayed towards these two propositions? It is not that C_4 has been subjected to rigorous empirical testing and found wanting, or that the potential for growth of social movements that embody the values of popular participation, and under what conditions, have been empirically investigated and found to be slight (chapter 11). I think that the source of the certitude is that the P-side is so deeply committed to the modern valuation of control and the values of capital and the market that the presuppositions of these values have been put outside the domain of scientific investigation.

8. Some adherents of the C-side (appendix: 1, variants, con) consider my viewpoint here much too modest, but I don't think that the empirical record (at the present time) supports the radical rejection of all uses of TGs.

9. A variety of additional questions, which I will not discuss, are relevant here (e.g.): Could TGs be "designed along agro-ecological principles"? (Welsh, et al. 2002). Might the time to consider TG innovations be after having dealt with land use and water problems in ecologically friendly ways? (Sanchez 2004).

10. When the range and plurality is determined (albeit in formally democratic societies) by the interests of the principal sources of funding for research, it is unlikely that neutrality will be furthered, and impartiality too—for it can be difficult not to reject a hypothesis, for example, C_4, which cannot be tested under the favored strategies, and for which the available evidence is not normally acknowledged in the mainstream, even though consequently the lack of

evidence for it may simply be an artifact of not enough relevant research having been conducted. Kitcher (2001), also concerned about questions of pluralism (though he casts them somewhat differently), speculates about what forms institutions for democratic determination of scientific agendas might take.

11. The presumption in favor of P_4a, so defended, might hold in some places (the United States) but not others (poor countries). Thompson himself appears to hold such a view. He certainly does not think that values like sustainability and empowerment of the poor should be routinely subordinated to the furthering human control over natural objects, especially if that control is mediated by market mechanisms (Thompson 1995; 1997; 2003a).

12. The P-side would, no doubt, expect high standards of "proof" to be invoked for the rebuttal of P_4a (see section 11.4).

Part III

PROLEGOMENON TO EMPIRICAL INVESTIGATION OF FUTURE SOCIAL POSSIBILITIES

Chapter Eleven

The Sociocultural Location of Alternatives to Transgenics

The successful conduct of research under agroecological strategies, the expansion and improvement of agroecological farming, and the activities and growth of movements that embody the values of popular participation are inseparably linked. The three flourish or decline together. If agroecology cannot be improved so that its methods become more widely adopted and so that it becomes a real alternative to TG-oriented farming, the explanation could be either that research conducted under agroecological strategies faces insuperable limits, or that there are irremovable obstacles to the growth of the movements. P_4 implies the former, P_4a the latter. Their combined force accounts for the sense of the necessity and inevitability that the P-side claims to accompany innovations of TGs.

The P-side claims that the authority of science gives support to this sense of necessity and inevitability. The thrust of the argument of part II is to cast doubt on this claim. It is true, of course, that TGs are the products of research conducted under materialist strategies, and that the efficacy of TG technology has been confirmed in the course of such research. P_4 is crucial to the case for legitimacy of using TGs on a large scale now, however, and its endorsement would only have strong empirical credentials if C_4 were subjected to investigation under a plurality of strategies, with the outcome that the promise represented in it were not vindicated. In the absence of that investigation, I traced endorsement of P_4 to endorsement of P_1 and commitment to the modern valuation of control. But that does not put the authority of science behind endorsing P_4. On the contrary, the aim of science (section 3.2), and commitment to impartiality and neutrality, point strongly to the need to conduct more research that investigates C_4.

Alternatively, if (instead of P_4) P_4a is endorsed, investigation of C_4 may appear to be simply irrelevant. But it would be irrelevant only if the grounds for endorsing P_4a do not include that P_4 has been reasonably endorsed. In section 10.6, I only indicated how the argument of the P-side is recast by replacing P_4 by P_4a. I did not consider the grounds for endorsing it. If it is endorsed principally because it is a presupposition of rationally holding the values of capital and the market (those of neoliberalism—section 6.3), then the legitimacy of using TGs on a large scale now would derive from a socioeconomic commitment, not from the authority of science. When the trajectory of capital and the market is considered inevitable (or beyond challenge), then the sense of the necessity and inevitability of TG innovations will appear to be part of the "common sense" of the times. However, should there be compelling empirical evidence to endorse P_4a, then that would enable the authority of science to be reclaimed by the P-side.

11.1 EMPIRICALLY INVESTIGATING "... IS NOT POSSIBLE"

In this chapter, I will begin to explore the question of how to investigate empirically propositions like P_4a. The issues are complex, and I do not pretend that this chapter is more than a prolegomenon to addressing them. I began the chapter with the statement: "The successful conduct of research under agroecological strategies, the expansion and improvement of agroecological farming, and the activities and growth of movements that embody the values of popular participation are inseparably linked." My goal is just to make clear how important this statement is, when we want to address the question: How to investigate empirically future social possibilities? That is because addressing it requires attention to the sociocultural location of alternative projects that contest predominant current social trajectories.

The various versions of the argument of the P-side build upon a series of claims of the form: " . . . are (is) not possible":

(1) Better forms of farming . . . are not possible. (P_4)
(2) Within the trajectory of the capital and the market, alternative forms of farming . . . are not possible. (first part of P_4a)
(3) Significantly furthering value-outlooks that do not contain the modern valuation of control is not possible. (Presupposition (d) of the modern valuation of control—section 1.1)
(4) Significantly furthering value-outlooks that do not contain the values of capital and the market is not possible. (This incorporates the second part of P_4a.)

These propositions, to a considerable extent, stand or fall together. Certainly, if there were strong evidence for endorsing (1)—that is, if, after conducting the relevant research, there were compelling evidence against C_4—this would provide evidence in favor not only of P_4, but also of the other propositions, and one might wonder whether there could be strong empirical support for endorsing (3) and (4) that does not draw upon independent evidence for endorsing (1) and related propositions pertaining to other key social practices. Endorsement of (3) might be derived from that of (4), in view of the mutually supporting interaction of the values of capital and the market and the modern valuation of control within predominant contemporary institutions, and so I will take (4) to be the key proposition.

Is there empirical support for (4) that does not depend upon independently endorsing P_4? If so, it would have to include rebuttal of evidence to the contrary put forward by movements that articulate, and aspire to embody to a greater degree, values such as those of popular participation (section 6.3).

The World Social Forum

The movements that constitute the World Social Forum (WSF) are a case in point (appendix: 1, sources, con). I discuss them, not because they are the only challengers to the trajectory of neoliberalism or because I think that they have clearly established themselves as the wave of the future, but because they take issue with (4) explicitly and directly, with awareness that their alternative claims need to be grounded in empirical investigations, and because their agricultural wing is in the forefront of agroecological developments. (WSF is an important part of the sociocultural location of agroecological research and development.) Clearly, given that the life of everyone depends on agricultural productivity, alternatives to the trajectory of capital and the market count on the grounds for endorsing C_4 (or a close variant of it) becoming strengthened in the course of further research of the kinds described in section 10.1.

"Another world is possible" is the challenging and unsettling signature of WSF. It is intended to flatly contradict (4); "another world," that is, a mode of structuring society which challenges the trajectory of the predominant forces of the current world (capital and the market), is possible. The signature statement is intended to express a factual judgment or, rather, a conjecture about a matter of fact for which a significant measure of empirical evidence is at hand and for which evidence mounts as, in practice, the movements are able to discern the contours of this other world. The other world becomes formed in anticipation of their organizations, and more developed as their effective collaboration with one another is established. Thus, "another world is possible" connotes the defiant rejection of a staple of the legitimacy of neoliberalism; it connotes an ethical critique of the actual social world whose structures (those of neoliberalism) are deemed defective as ideal and repugnant in actualization, structures from whose effects emancipation is desired; and it connotes an imperative to act so that this other world will become actualized. Not only is another world possible, it is desirable. Simultaneously, then, the statement is intended to express a partially confirmed conjecture, and a desire, a hope, an aspiration, an objective, and a commitment to act. If "another world is possible" does become well confirmed, that will be only as a causal consequence of the practices of those who succeed in actualizing this other world. The clinching case for a possibility is its actualization—but that does not mean that, prior to its actualization, there is no evidence supporting endorsing that it is possible or that not having been actualized is compelling evidence for its impossibility. Any actual phenomenon in the world today (if there are any) whose dynamic trajectory is towards its actualization constitutes evidence to endorse that another world is possible.

I will apply the conclusions reached in part I to the discussion of questions about the empirical status of "another world is possible," the grounds for endorsing or not endorsing it, the way in which fact and value come together in it, and the interplay of factual and value judgments in the types of empirical inquiries that are pertinent to its cognitive appraisals.[1] Some of the key questions will be: What is, or ought to be, the relationship between investigation in the social sciences (and the sciences in general) and the practices of actual emancipatory movements—both their social and political organizing and their participation in efforts to develop alternative approaches in areas like agriculture? How, under what strategies, should research be conducted in the social sciences—in a disciplined, systematic, empirical way—so as to be of service to these movements: to inform their practices by providing knowledge about means to their ends, about impediments that they face, and about what the realm of possibilities open to them may be?[2] How does investigation of (and participation in) the emancipatory possibilities represented by these movements (and the practices that might lead to their development) contribute towards furthering the aims of the social sciences?

These questions are important for movements, like those of WSF, that characterize themselves as emancipatory.[3] Their pertinence is supported by my conclusions (part I) that science is not a unitary activity and that it is (cognitively) legitimate to engage in research conducted under strategies (e.g. agroecological strategies) that are identified as those with the potential to identify possibilities of interest to those who hold the values of popular participation (section 6.3). What strategies need to be adopted in order to investigate proposition (4) and its negation, "Another world is possible"? To dismiss the admissibility of research conducted under these strategies, prior to sufficient research having been conducted so that their fruitfulness can be tested, is de facto accepting (4), but accepting it in violation of impartiality. Those who seek to specify strategies that are responsive to the interests of the values of popular participation, and so with the capacity to investigate "another world is possible," should not, under present conditions, concede the cognitive (scientific) high ground exclusively to mainstream science and its dominance by materialist strategies (and their counterparts in the social sciences).

I turn now to a discussion of emancipation and emancipatory movements.

11.2 EMANCIPATORY MOVEMENTS

A fundamental dimension of human well-being is the exercise of cultivated, effective agency (*SVF*: chapter 9), which is exercised when people act in the important aspects of their lives, informed by their beliefs, so that desires expressing the values they hold are regularly satisfied. One holds something as an ethical value, for example, if one considers it a characteristic of a fulfilled, flourishing, meaningful, or well-lived life or a relation among people that fosters (and partly constitutes) fulfilled lives, and if it is partly constitutive of one's personal identity (section 3.1; section 6.3; *SVF*: chapter 2). Values encompass both needs—necessary conditions for a life worthy of a human being; and wants—expressions of one's personal (social and cultural) identity. Effective agency rests upon a variety of conditions, for example: that normally the beliefs that inform one's actions are true (cf. Bhaskar 1986: 170); that it is possible to manifest one's values more fully under prevailing conditions; that one has appropriate bodily capacities, skills, and motivations (cf. Bhaskar 1993: 280); that one can count on others to interact in appropriate ways with one's actions; and that one has access to and control over relevant material (technological, biological) objects.

When the conditions for effective agency do not obtain, agency is diminished. Diminished agency may be (and often is) a significant social phenomenon. Then, we would expect it to be accompanied by (and causally linked with) diminishments of other dimensions of human well-being: the bodily dimension, reflected in such sufferings as sickness, hunger, and malnutrition; and the social dimension, reflected in entrenched human rights abuses and such sufferings as loneliness, exclusion, abandonment, and alienation. We would expect these sufferings to be accompaniments of diminished agency, for characteristically, when free to be so, agency is exercised to further well-being in all of its dimensions, and so to prevent or cure such things. Moreover, diminished agency involves its own characteristic sufferings: a sense of meaninglessness, frustration and incompetence, and depression; and the diminishment of all dimensions of well-being together can generate all sorts of psychological pathologies. I will call the phenomenon of diminished agency together with its attendant sufferings "oppression." Those who experience oppression, which is present when the conditions of effective agency are absent or difficult to access, never desire it. It may be accepted with resignation, or experienced as a state from which one wants and needs to be rid—as a state from which one desires emancipation.

Emancipation and Emancipatory Activity

Emancipation is becoming rid of the state of oppression and gaining the conditions for effective agency.[4] Furthermore, to get rid of oppression requires getting rid of the causes of the absence of the conditions of effective agency. These causes are many and varied, and they operate through a variety of interacting mechanisms. Where oppression is a significant social phenomenon the causes may include social structures, which do not serve to provide the conditions for effective agency for all their participants. Then emancipation would require the transformation of these structures, but transforming them per se is not sufficient for emancipation, for structural change may (and often does) generate yet other forms of oppression. Structural transformation serves emancipation only if it produces structures that generate conditions for effective agency, and thus structures that embody highly values held (or that come to be freely held) by those desiring emancipation). Where a social structure is a major causal factor in maintaining oppression, how is emancipation possible? Let us first consider briefly the relationship between social structures and personal action.[5]

Social Structures

A social structure (SS) is a set of more or less enduring relations among its participants that define roles or places for its participants' activity. (All symbols used are listed in the glossary.) SS is reproduced by the actions of agents (participants in the structures) carrying out their roles, typically as side effects of intentional action of which the agents may be unaware. But SS can also be modified and transformed by intentional activity. SS embodies specific personal and ethical values by providing roles in which they can be manifested highly, by providing the conditions (material, financial, etc.) for their high manifestation in many domains of a participant's life and, in some cases, by supporting institutions that suppress efforts to act in accord with other values. In doing so, SS manifests (to a greater or lesser degree) certain social values depending on how its relations are causally linked with the distribution of wealth, power, access to scientific knowledge, respect, etc. There is mutual dependence between, on the one hand, the stability of SS over time and, on the other hand, the personal (and ethical) and these social values being held (or at least not significantly contested) widely among its participants. SS provides the conditions for the high manifestation of these values, and their manifestation and personalization in action contribute to reproduce SS.

The values that agents hold have factual presuppositions. The most important of these are about the character of human well-being (human nature), and about the possibilities open to action—including the possibilities for the greater manifestation of the values held by the agents, as well as of competing values, within SS (see section 3.1: *SVF*: chapter 2). Holding values presupposes having specific beliefs, and where those values are held we would expect the beliefs to be nurtured. (Cf. the account of the presuppositions of the modern valuation of control, section 1.1.) There is also, therefore, mutual dependence between the stability of SS and (relevant) participants holding certain beliefs, including some about SS itself. Social structures can be significant factors that cause certain beliefs about themselves to be widely held among their participants, where the "mechanism" of causation is that of the mutual dependence just referred to. In other words, a belief about SS may be held because its being held among relevant participants of SS is a necessary condition for the stability (legitimacy, hegemony) of SS, and agents desire (or do not contest) the stability of SS—for some, because they hold the values that can be highly manifested in SS; for others, because they discern no (acceptable) path to an alternative, or because the question of an alternative has simply not arisen to their consciousness or been effectively suppressed. Under appropriate cultivation, such beliefs may appear to be part of the prevailing "common sense" of the times of a culture, just as endorsing the core presuppositions of the modern valuation of control is (as illustrated in my account of the P-side in part II) part of the "common sense" of many contemporary scientific institutions and their spokespersons.

Possibilities Left Open in Dominant Social Structures

Nevertheless, SS need not close off all possibilities for acting according to values that it does not embody. By providing the conditions for its own reproduction through the agency of individual persons, at the same time, it provides conditions (and permits space) for some competing projects. Furthermore SS does not penetrate completely into all dimensions of life; within its confines there are institutions (e.g., family, religious, cultural, educational, political party, local governments, grassroots movements, scientific research) in which other values may come to the fore, at least in articulation. It is in the spaces left open by SS, and in the institutions whose values are in tension with those embodied in SS, that the possibilities of emancipation can arise.

Emancipation, remember, is from oppression, that is, from diminished agency and its attendant sufferings that may be co-caused by actual social structures, and for the exercise of effective agency. Oppression, I said, is never a desired state. Many oppressed people accept this state with resignation, for they have become convinced that any attempts to move toward emancipation will come to nothing. But others find it so intolerable that they will actively take steps that they discern might enable movement toward emancipation and, in doing so, effective agency is already exercised—to whatever limited degree it may be. It cannot be acting alone, for an individual lacks the power to bring about conditions for effective agency, although an individual can inspire others to act. It must be collaborative action directed toward, for example, alleviating some of the sufferings that constitute the oppressive state that is experienced. The activity will manifest such values as solidarity and commitment to an array of social/economic/cultural rights (the values of popular participation, V_{pp}—section 6.3), the values to be highly embodied eventually (it will be anticipated) in changed social structures, and the collaboration involved will typically have roles for the participation of people who are not members of the oppressed groups.

Summing up, I will say that emancipatory activity is collaborative activity, engaged in by people who experience oppression and by those in solidarity with them, that is expressive of their effective agency that aims both to alleviate the sufferings being experienced and to create conditions for the effective agency of everyone—where the values, manifested in the activity and embodied in the movements and institutions that encompass it, anticipate the values desired to be embodied in transformed social structures.

The Development of Emancipatory Activity

Emancipatory activity does not depend on antecedently changing SS; it can occur in the spaces left open in SS and even draw upon conditions made available within it. It involves discerning possibilities for the exercise of effective agency under conditions that generate diminished agency in many ways. The affirmation that there are such possibilities is demonstrable only with their actualization. (Recall: "Another world is possible.") Once demonstrated, their extent remains open to further demonstration in practice. Fuller emancipation builds upon finding space for the exercise of effective agency, and then expanding it through an evolving practice. Only through such a practice can the possibilities of the emancipation of all oppressed groups (with their great variety of values and interests) be tested. Note, however, that even initial successes of emancipatory activity begin to change SS by modifying some of its constitutive relations—for relations, which displace those SS would maintain among its occupants, are created among those who cooperate in the practice.

These newly created relations, and the values manifested and embodied where they come into place, anticipate the structures that would have to come to be to provide conditions for the effective agency of everyone, and they begin to suggest a "horizon" against which those seeking emancipation grasp what emancipated structures would be like (Ellacuría 1991). It follows that, unless there is careful coordination among those engaged in these practices and those involved in political (or, sometimes, military) activity aiming directly to change the fundamental relations of SS, it is unlikely that the latter activity will further the objectives of emancipation. Since emancipation concerns the exercise of cultivated, effective agency, it must grow out of the emancipatory activities of the oppressed themselves, and cannot be fitted to structural objectives antecedently designed by intellectuals or politicians.

The oppressed are the primary agents of their own emancipation. Then, for its completion, emancipation grows from the collaboration of ever-larger groups; and, if it is indeed to be all-inclusive, it will need structures (SS') within which many of the values much prized by the privileged within SS can continue to be manifested. This does not mean that SS' will provide space for the high manifestation of all the values that are highly manifested and embodied in SS. SS and SS' cannot coexist with the same participants. (Not all genuine possibilities can be co-actualized.) But freedom does not require that people be able to shape their lives so as to manifest any values that they choose; it is consistent with freedom to restrict legitimate values to those whose manifestation is consistent with respecting the rights of others. By hypothesis, SS does not provide space for the manifestation of values concerning basic human rights that the oppressed hold. SS' needs to provide space where everyone can manifest values, freely chosen in the light of considerations about the rights of others, where lives into which these values are woven can be experienced as lives of well-being.

What values may be adopted, consistent with everyone being able (ceteris paribus) to manifest the values they come to hold, is clarified, shaped, and transformed in the course of the emancipatory practices (Lacey and Schwartz 1996), which may be expected to begin to actualize hitherto unrecognized personal and social possibilities. The movement toward SS', if these structures are to be adequate for the emancipation of everyone, must be sensitive to this point. While this represents an ideal, it is an ideal against which manifestations of the social value, democracy, should be evaluated. It also leaves open—to be settled as the emancipatory process unfolds—how radical is the structural change needed for emancipation. Nevertheless, reforms that might alleviate the suffering of the oppressed (even to a considerable extent) are not enough; the test is whether the conditions for the exercise of cultivated, effective agency become actualized.[6] Since effective agency presupposes certain conditions with structural (co-)causes, emancipation requires structures in which democracy is highly manifested—structures in which (in principle) everyone is able to participate in decisions that have impact on the availability of the conditions for effective agency; these will include decisions pertaining to such matters as the production and distribution of goods (manufactured and agricultural) and services, the goals and processes of the workplace, and priorities among (and the development of institutions to support) the array of social/economic/cultural rights in proper balance with civil/political rights.

11.3 THE MOVEMENTS OF THE WORLD SOCIAL FORUM AND OPPOSITION TO NEOLIBERALISM

Many kinds of oppression are experienced throughout the world today, and they have varied historical roots, including colonialism, dictatorships, fundamentalisms, racism, and sexism. They are maintained (reshaped and, for many, deepened) currently within the structures of neoliberalism (SS_{NL}) that, under the mantle of "globalization," are increasingly absorbing more and more economic, social, and cultural space throughout the world and penetrating into ever more domains of life and nature. The movements of WSF seek emancipation from diminutions of agency and sufferings that they diagnose as maintained and partly caused by SS_{NL}.[7] See section 6.3 for a brief account of neoliberalism, its predominant institutions, policies, and its values, those of capital and the market (label $V_{C\&M}$). Recall that $V_{C\&M}$ includes the modern valuation of control (V_{MC}). SS_{NL} manifests V_{MC} highly; it the foremost bearer of V_{MC} in the world today, and the further manifestation of $V_{C\&M}$ is implicated in the continued predominance of R & D conducted under materialist strategies (S_M) and successful application of techno-scientific discoveries. Today the legitimacy of SS_{NL} and its expansionary thrust are widely—though not without (perhaps growing) contestation—taken for granted, most emphatically within the powerful and influential institutions of many countries, but the sense of its inevitability (if not legitimacy) often extends down to the grassroots especially in advanced industrial countries.

The "Common Sense" of Neoliberalism

There can be mutual dependence (I said above) between the stability of a social structure and the acceptance among its participants of certain beliefs about it. Among the beliefs (the "common sense" of neoliberalism) on which the stability of SS_{NL} depends are (I suggest) the following: (i) all (or most of) the presuppositions of V_{MC} (listed in section 1.1), reflecting that SS_{NL} is the foremost bearer of V_{MC} today; (ii) widely shared beliefs endorsing individualist views of human nature, which emphasize individual agency and the individual body and de-emphasize the social character of human beings and their relationships to cultures and groups, human beings as choosers, centers of creative expression, consumers, who maximize "preferences" or "utilities" and the like—then, human well-being consists of (along with dimensions of bodily and psychological health) the actualized capability, which is enhanced by expanding human capabilities to exercise control over natural objects, to express a variety of ("authentic" or self-chosen) egoistic values (Lacey and Schwartz 1996). Finally, functionally interconnected with (i) and (ii) and among themselves, (iii), a set of propositions that includes the following:[8]

(a) In the foreseeable future there are no significant viable possibilities, for the high manifestation and embodiment of values, outside of SS_{NL}, and thus of values that are incompatible with $V_{C\&M}$. (I will abbreviate this, highlighting the antagonism of WSF: "Another world is not possible!")

(b) Within SS_{NL}, and its component institutions, there are continually expanding opportunities—eventually, in principle, available to everyone—for indi-

vidual people to manifest values that are self-chosen and that express their personhood authentically, provided that they take the appropriate individual initiatives. ("The neoliberal world is a "good enough" world!)
(c) Although significant negativities (sufferings such as poverty, hunger, disease, etc.) persist under SS_{NL}, and numerous people are not able to live in ways that could plausibly be considered to be expressive of their authentically self-chosen values and thus of a sense of well-being, nevertheless these phenomena have their principal causal origin in the previous structures that SS_{NL} has replaced (or is replacing), or in individual deficiencies, and not in SS_{NL} itself. ("Neoliberalism is not to blame!")
(d) SS_{NL} depends upon, and is a source of the maintenance and spread of democracy; more generally, democracy flourishes in an exemplary way under SS_{NL}. ("We all "really" want a neoliberal world!")

Proposition (a) is identical to (4) (in section 11.1), which incorporates the last part of P_4a (section 10.6). The other propositions, too, often have ripples in the controversy about TGs.

Questioning the "Common Sense" of Neoliberalism

The acceptance of most of (a)–(d) by key participants of SS_{NL} is (I suggest) essential for the stability of SS_{NL}. Yet, there is a striking disparity between them and the judgments made by the movements of WSF about the causes of the oppression they experience. Consider, for example, (c), "Neoliberalism is not to blame!" What the proponents of neoliberalism identify as negativities, inherited from the past, which can be addressed within SS_{NL} (as in proposals to address malnutrition with a techno-scientific solution—section 8.3), WSF identifies as oppression maintained and partly caused by SS_{NL}. As SS_{NL} become increasingly encompassing, the WSF points out, in ways that can be traced to shifts and implementations of its policies, there has been a growing gap between the handful of the rich and the vast majority of the poor, crippling and deepening national indebtedness in most poor countries (WSF, 2002a, item 13) and loss of national sovereignty as policies are imposed by international institutions (item 16), increased impoverishment and social disruption, selling off of national patrimonies to foreign owners, decreased social services and educational opportunities for the poor, weakening of educational and health systems, unstable (and often unhealthy or dangerous) conditions of employment and underpaid work, access to land for small-scale farming denied, the devastation of hope and the accompanying rise of criminal and terrorist violence, the exposure of the poor periodically to especially intense sufferings induced by fluctuations of the global economy (or natural disasters), increased pressures on the environment, and growing cultural homogenization.[9] (Not only is neoliberalism to be blamed, a neoliberal world is far from good enough.) These phenomena ensure that many (especially poor) people are excluded now from experiencing the kind of value furthered by SS_{NL}, let alone value that they might prioritize in the light of their cultural traditions. Unless, then, there is a reversal of trends it follows that, even if SS_{NL} does expand opportunities, (b), as well as (c) and therefore also (d), is false.[10] If a decisive case could be made for (a), one might then be able to also defend (b) and (c)—but, without a strain, not (d)—despite the miserable reality described by WSF. However, it would not be a defense accompanied by celebration, but by tragic resignation. Proposition (a) is the most fundamental one of the set.

The (Social) Science That Needs to Inform Emancipatory Movements

It is crucial to the prospects of the movements of WSF that the analysis expressed in the previous paragraph (or better analyses) about the causal role of SS_{NL} in generating the listed social pathologies be soundly based empirically. More generally, emancipatory practices need to be informed by scientific (systematic, empirical) understanding, for effective practice is likely to be impaired if it is not informed by sound understanding of matters such as the following: (i) Causes of the condition experienced as oppressive (and of the nature, variation, and variety of its attendant sufferings), including structural causes; (ii) sources (and causes of emergence) of emancipatory possibilities, means to advance their actualization, and structural conditions for the effective agency of those with the values embodied in emancipatory movements; (iii) unintended consequences and unacknowledged conditions of action; (iv) potential constraints on and obstacles to emancipatory activity—including mechanisms whereby the beneficiaries of dominant structures act to close the spaces that are opened for emancipatory activity; (v) spaces in actual structures where alternatives might emerge; (vi) available evidence for the presuppositions of the values manifested and embodied in actual structures, especially those that are inconsistent with the possibility of emancipation; and (vii) how to conduct (e.g., agricultural, communicative, and educational) practices that may provide material, ecological, and social conditions for the growth of emancipatory practices.

The oppressed (those among them desiring emancipation) have an interest in inquiry that may produce understanding of such matters, and in access to its results, so as to be able to act in the light of sound understanding (including self-understanding).[11] They must also be active agents in carrying out the relevant investigations. They have an indispensable epistemic role in them, because the experience and testimony of the oppressed is essential (though not incontestable) for accurate charting of their sufferings, of many of the subtle effects of activity carried out within dominant structures, of where there are possible spaces for emancipatory action, of the sources of alternative values, and—above all—of what are the authentic values of the oppressed and those struggling for emancipation.[12] Without adequate evidence pertaining to such matters, the character and the effects of the structural (co-)causes of oppression cannot be analyzed scientifically, and the possibilities for emancipation cannot begin to be identified with any confidence. The research that can inform emancipatory practices, then, requires the collaboration (and, to some extent, overlap) of researchers and agents aspiring for emancipation (Bhaskar 1986: 200–211; Lacey 2000b; 2003c; Santos 2002). The values involved in such research include solidarity with those who are suffering from current policies and a sense of justice that leads to indignant responses where policies are implemented without heed for the suffering they cause for the poor. The research strategy deployed does not admit subordinating the question: "How to meet the needs of the poor, how to redress their sufferings, how to bring about conditions in which the poor can manifest values that are genuinely their own?" to: "How to design and implement market and other reforms favored by neoliberalism?" or "How to solve the problems of the poor with techno-scientific innovations?"

11.4 A QUESTION FOR SCIENTIFIC INVESTIGATION: IS ANOTHER WORLD POSSIBLE?

Unable to share in the value available within SS_{NL}, the oppressed—moving beyond "simple moral protest" (WSF 2002b)—may struggle toward emancipation as they discover, create, and develop alternative spaces in which they can manifest their own values: "Spaces of democratic participation, non-capitalistic production of goods and services, creation of emancipatory knowledge,[13] postcolonial cultural exchanges, new international solidarities" (Santos 2002).[14] These are spaces that manifest, for example, V_{PP}, the values that are held by many (though not all—there is ongoing dispute within WSF about what the values of "another world" should be) of the movements within WSF. When WSF affirms that another world is possible, it is affirming that it is possible for these alternative spaces to become sources of structures that will come to manifest and embody V_{PP} highly. This conflicts with the belief (a) (previous section); and so the signature statement of WSF contradicts a staple of the common sense of neoliberalism. The conflict is about a matter of fact. What kind of scientific (systematic, empirical) inquiry might potentially resolve this conflict? Under what strategy (strategies) must it be conducted? What kinds of empirical data are needed to appraise affirmations: another world is—or is not—possible.

Adopting a strategy involves specifying the kinds of possibilities that can be investigated under it, in important cases possibilities considered to have special social value (section 1.3). Those who hold $V_{C\&M}$, and believe the presuppositions, (a)–(d), tend to adopt strategies that limit the possibilities that may be investigated largely to those that could be actualized under actual hegemonic social structures. These may be the most common type of strategies adopted in contemporary social science (I will label them S_{CSS}). Research conducted under S_{CSS} cannot provide evidence for or against "another world is possible," since it does not investigate what may be possible under novel structures. I have indicated that a (co-)cause of (a), "another world is not possible," being widely believed is its role as a necessary condition for the stability of S_{NL}. Yet, as a factual proposition, (a) is open to the deliverances of evidence and argument and, were the relevant research conducted, the evidence might in fact support its negation, "another world is possible." Thus, research that confirmed the latter would have causal impact on SS_{NL}; it would contribute towards undermining a necessary condition for its stability. This may explain why those who value the stability of SS_{NL} (or, more exactly, who value the surety that the trajectory of SS_{NL} will continue to augment the manifestation of $V_{C\&M}$) tend to withhold material and other support and prestige from research practices that might lead to the confirmation of "another world is possible." This is not conspiracy: a strategy is adopted (in part) so as to enable the identification of valued possibilities (section 1.3); the fact that it does not identify nonvalued or negatively valued ones may be seen simply as an unavoidable by-product. (Like research that is framed by P_1 in part II, however, it is in tension with impartiality and neutrality.) Then, the status of (a) is simply that of "untested consciousness," associated with valuing the stability of SS_{NL} and, when untested consciousness remains actually not contested (as it tends not to be under a stable SS_{NL}), the absence of credible contestation may come to be portrayed as positive evidence for endorsing (a). By way of this mechanism, (a) comes to function as part of the prevailing "common sense" within SS_{NL}. It follows that a research strategy under which "another world is possible" might become a reasonably endorsed hypothesis is only likely to be adopted where the values which are manifested in SS_{NL} are contested and alternative strategies are adopted in research.

Strategies for Investigating "Another World Is Possible"

In order to investigate "another world is possible" empirically, and, therefore, its negation (a), one would have to adopt strategies (S_{SS}) that investigate social structures, their generative powers (concerning negativities as well as concerning the enhancing of the values manifested and embodied within actual structures), the patterns of their historical causation, and their potential to change (Lacey 1997: 214–16). These are the same strategies that are needed to address the questions raised in the last subsection of section 11.2, and research conducted under them depends on the collaboration in solidarity between researchers and agents in the emancipatory movements. These strategies also need to have the conceptual resources to chart observationally the unfolding (and setbacks) of the movements, and thus to identify and describe actual phenomena (if there are any) whose actual trajectory is toward the actualization of structures that manifest and embody V_{PP}. This is important, for the only relevant empirical evidence that such structures are possible (in the absence of their full actualization) is bringing into being phenomena with such a trajectory.[15] Only the successes of emancipatory movements and their unfolding developments provide relevant empirical evidence pertinent to the possibilities of emancipation; only in actual practice can the genuineness of emancipatory possibilities (as distinct from the desirability of there being genuine emancipatory possibilities) be empirically confirmed.[16]

Adopting S_{SS} bears mutually reinforcing relations to holding V_{PP}. One of the presuppositions of holding V_{PP}, "another world is possible," can be investigated under S_{SS}, and the degree of support for its endorsement is deeply intertwined with the degree of manifestation of V_{PP} in the movements.[17] At the same time, if evidence supports endorsing "another world is possible," it counts against endorsing (a); and so at the same time that it supports a presupposition of holding V_{PP}, it undermines to some extent a presupposition of holding $V_{C\&M}$. Questions about the adequacy and sufficiency of evidence (like those discussed in section 9.4; section 9.6) will arise here. Clearly for those who hold $V_{C\&M}$ the standards (*SVF*: 62–66) of adequacy will be more demanding—but, if they are not to put (a) outside of the realm of possible empirical appraisal, it is incumbent on them to identify the standards that they consider appropriate (section 9.4; Lacey 1997: 222-23). Despite the difficulties of interpretation, the outcome of social science research may have impact in the realm of values—by either consolidating or undermining the presuppositions of value-outlooks.

The Value of Neutrality

I have already indicated that when theories are produced under a strategy, the adoption of which bears mutually reinforcing relations with holding specified values, on application they are likely to favor especially projects that serve interests associated with these values. On the whole, we should not expect neutrality to be achieved from research conducted exclusively under one kind of strategy. Research conducted exclusively, in the natural sciences, under S_M or, in the social sciences, under S_{CSS}—strategies held (de facto) to consistency with key presuppositions of neoliberalism—does not in general produce neutral products.[18] Therefore, there can be no objection in principle, based on cognitive considerations, to conducting research under strategies (e.g., S_{SS} or S_{AE}), the adoption of which bears mutually reinforcing relations with holding V_{PP}. Its products, too, are not neutral. In all of these cases we have instances of theories, whose endorsement (or even, when appropriate, acceptance in accordance with impartiality) does not generate neutrality.

Throughout part I, I argued that neutrality could not be furthered from research conducted under a single strategy. Neutrality, where products of scientific activity are available in principle to serve more or less evenhandedly projects of interest to all actually competing value-outlooks, is only approachable if scientific institutions as a whole support research conducted under a plurality of strategies—in part II, I argued that agroecological strategies should be among the plurality and, in this chapter, I have sketched a case for including S_{SS}. To reach this conclusion, those who hold V_{PP} need not (and should not attempt to) de-legitimate research conducted under other strategies, certainly not the credentials of any theories that have been accepted in accordance with impartiality, even if accepted theories may make it necessary for them to reorganize priorities.

Neutrality cuts deeper still. A strategy may be adopted, I have maintained, in large part because of the mutually reinforcing relations adopting it has with holding a value-outlook; and the value-outlook (in turn) has presuppositions of a broadly factual kind, and it usually also plays a role in supporting the legitimacy of applications of theories developed under the strategy (e.g., the role of the modern valuation of control in legitimating widespread and expanding uses of TGs at the present time—part II). The evenhandedness that is part of the ideal of neutrality thus requires that these presuppositions be submitted to empirical inquiry—that not only the knowledge that informs the efficacy of applications be accepted in accordance with impartiality, but also the presuppositions of their legitimacy (where they cannot at present be accepted in accordance with impartiality) be endorsed only in the light of adequate and sufficient evidence (chapter 9; chapter 10). Even though it is not especially well manifested in actual scientific practice, neutrality is a value that should guide scientific practice; it is a value built into the very aim of scientific activity (section 3.2). Aspiring to further the manifestation of neutrality in scientific practices requires that a plurality of strategies should be in play. It also requires that all factual presuppositions of the value-outlooks linked with the adoption of strategies, and the legitimacy of applications, become objects of scientific (systematic, empirical) inquiry. Not only is this a value of scientific practice, but also it serves actual emancipatory interests by requiring that presuppositions of the legitimacy of SS_{NL} become objects of empirical inquiry; and that cannot happen, I have argued, unless the emancipatory movements themselves become a key object of inquiry in research conducted under S_{SS}.

NOTES

1. The argument of this chapter has been greatly influenced by my interacting for many years with writings of Roy Bhaskar (1986; 1993; 1998). For details (and numerous references to passages of Bhaskar's works and their affinities with ideas of WSF) see Lacey (2002b), from which most of this chapter is drawn (see also Lacey 1997).

2. This question parallels the one considered in section 10: What sort of research needs to be conducted, if further evidence is to be provided for C_4? That research is not guaranteed to come out in favor of C_4; if research conducted under the relevant strategies turns out not to be fruitful that will become a ground for endorsing P_4. Similarly, here, if the research project indicated turns out not to be fruitful, that would provide a good reason to endorse (4).

3. In presenting the position of movements of WSF, I use their own terminology, and hence such terms as "emancipation" and "oppression." Obviously these terms are value laden and only those who hold values like those of popular participation will use them for descriptive ends. They are examples of "thick ethical terms" (Williams 1985). Sentences that contain them may be used to make both value judgments and factual assertions. (The value judgments they make are markedly less compelling if the corresponding factual assertions fail to gain empirical support.) That does not prevent the factual assertions they make—like those made by value-assessing statements (section 3.1), which also use value-laden terms—from being appraised in terms of available empirical data and the cognitive values. I recognize that a number of assumptions are implicitly made here, for which there is no space to defend here. This is a symptom of the fact that this chapter is only a prolegomenon.

4. My account of "emancipation" (or "liberation") draws upon Latin American liberation theology (Lacey 1985; Ellacuría 1991), Bhaskar's writings, documents of WSF (2001; 2002a; 2002b), and intellectuals associated with WSF (Santos 2002; 2005). All the cited works emphasize that the actual and its predominant tendencies do not determine, though they set limits to, what is possible.

5. Here I make use of (with minor, mainly terminological, modifications) Bhaskar's analysis of social structure and also his "transformational account of social action" (Bhaskar 1998).

6. The tension between those who favor agroecology and those who want to use TGs in programs intended to address needs of the poor reflects this issue. Strengthening agroecological practices is an important project to those who hold the values of popular participation, and that is why weakening them conflicts with democratic ideals (section 10.3).

7. In its Charter of Principles, WSF (2001) defines itself as "opposed to neoliberalism and to domination of the world by capital and any form of imperialism . . . to a process of globalization commanded by the large multinational corporations and by the governments and international institutions at the service of those corporations' interests, with the complicity of national governments."

8. This list was made in the course of reflecting on what would count as plausible premises of an argument that would justify valuing SS_{NL} highly, or regarding it as legitimate. Versions of the propositions are stated frequently in the press, and they arise in discussions with supporters of SS_{NL}, who often respond to critics—with a tone of exasperation—citing these propositions as if they were part of common sense. Detailed empirical investigation of the following questions would be very useful. (i) Are the propositions I identify actually widely believed? If so, how are they articulated, and what evidence is cited in support of them? If they are not widely believed, how do people actually think that $V_{C\&M}$ are legitimated? And are they committed to $V_{C\&M}$ or are they simply resigned to the power exercised by the predominant institutions of SS_{NL}? (ii) What are the social mechanisms—educational, press, political, and business rhetoric, etc.—that foster belief in these propositions and (if I am right) render them part of the "common sense" that reigns wherever SS_{NL} prevail? (Lacey 1997: 238–39). I have already indicated (section 10.6) that I think that the sense of the inevitability of the trajectory of SS_{NL} provides ready ground for legitimating the widespread use of TGs at the present time.

9. "This system produces a daily drama of women, children, and the elderly dying because of hunger, lack of health care, and preventable diseases, families are forced to leave their

homes because of wars, the impact of large-scale development projects, landlessness and environmental disasters, unemployment, attacks on public services and the destruction of social solidarity. . . . The neoliberal economic model is destroying the rights, living conditions and livelihoods of people. Using every means to protect the value of their shares, multinational companies lay off workers, slash wages and close factories. . . . Governments faced with this economic crisis respond by privatizing, cutting social sector expenditures and permanently reducing workers' rights. The recession exposes the fact that the neoliberal promise of growth and prosperity is a lie. . . . Neoliberal policies create tremendous misery and insecurity. They have dramatically increased the trafficking and sexual exploitation of women and children. Poverty and insecurity creates millions of migrants who are denied their dignity, freedom and rights" (WSF 2002a). Also: this system is "incapable of responding to the primary function of the economy: providing the basis of physical and cultural life for all humans on the planet" (WSF 2002b). Obviously, substantial social scientific work is needed to establish that there are causal relations between these phenomena and SS_{NL}. Looking into such causal relations—the causes of current social pathologies—is typical of social science that would serve emancipatory movements.

10. I recognize, of course, that these summary remarks are far from the end of the discussion on the matter. They suffice for present purposes, which is to indicate the sorts of issues that need to be investigated—either to confirm the claims of neoliberalism or those of their WSF antagonists. WSF is pointing accurately to realities of the contemporary world. Questions about their causation and solution remain open to inquiry—and urgent practical proposals for redressing them.

11. WSF (2001) considers itself as "a movement of ideas that prompt reflection, and the readily available circulation of the results of that reflection, on the mechanisms and instruments of domination by capital, on means and actions to resist and overcome that domination, and on the alternatives proposed to solve the problems of exclusion and social inequality that the process of capitalist globalization, with its racist, sexist, and environmentally destructive dimensions, is creating internationally and within countries."

12. I have already mentioned that the values of the oppressed may be clarified, shaped and transformed in the course of emancipatory struggle. Scientific knowledge may play a part in this by, for example, confirming or disconfirming presuppositions of the value-outlooks held. But the authentic values of those engaged in emancipatory struggle reflect their own desires and the outcomes of their deliberations. These cannot be derived from any scientific investigations. Intellectuals face the temptation of thinking that they can grasp (from their analyses) what the poor need for their emancipation, and not to pay adequate heed to what the poor say they need and foresee as likely effects of "well-intentioned" proposals. Giving in to this temptation is to continue the "management" of the poor, which is one of the structural phenomena that impede the actualization of the possibilities of emancipation.

13. My discussions of agroecology (section 5.4; section 8.3; chapter 10) and Shiva's ideas (section 7.2) exemplify this point.

14. For details on the various aspects of the spaces referred to in this quotation and supporting quotations from WSF documents, see the longer version of my article, Lacey (2002b), available at my website, www.swarthmore.edu/Humanities/hlacey1/.

15. Compare WSF (2002b: final paragraph). Evidence that steps are taken within dominant structures to prevent alternative possibilities from emerging or developing is not evidence that alternative structures (with the specified characteristics) are not possible. Neither is showing that (a) is held causally because of its role in maintaining the stability of S_{NL}. Similar points were made in the discussions of risks of using TGs (chapter 9) and alternatives (chapter 10).

16. Since such successes depend (in part) on appropriate knowledge being available, the conduct of research in the (social) sciences is an important factor. Solidarity with those seeking emancipation is, thus, a key virtue of the investigators. Also, "The simple addition of these efforts will not be able to give birth to the necessary transformations. The constitution of a force that is truly organic will need a vision of the whole, that is built on a day-by-day analysis. Fruit of a continual interaction between action and knowledge, enriched by the social and cultural experience of all the people of the world, it will necessarily call for the privileged contribution of the immediate victims of contemporary capitalism and of all the discriminations that it accentuates: the impoverished peasants, industrial workers, women, the unemployed, the urban poor, indigenous people, youth without a future . . . " (WSF 2002b). "None of these thematic initiatives taken separately will succeed in bringing about counter-hegemonic globalization. To be successful their emancipatory concerns must undergo translation and networking, expanding in ever more socially hybrid but politically focused movements. In a nutshell what is at stake . . . is the reinvention of the state and of civil society. . . . This is to be accomplished through the proliferation of local/global public spheres in which nation-states are important partners but not exclusive dispensers of either legitimacy or hegemony" (Santos 2002); compare WSF (2001: item 14).

17. One might adopt S_{SS} in the first instance simply in order to put (a) under empirical investigation; but, in view of the collaboration between researcher and agents in the movements required for the conduct of the research, it is hard to see the research being conducted by researchers who do not hold V_{PP}. This imposes no barrier per se to obtaining results that accord with impartiality, though it may to those who do not share these values being able to recognize such results. (This reinforces the arguments for strategic pluralism that are interspersed throughout this book.)

18. I have already indicated that a measure of neutrality is present with some (even many) results obtained under S_M—see, for example, the discussion of agroecological strategies and of how they make use of knowledge gained under S_M. That does not add up to the "evenhandedness" required of neutrality.

Appendix

1. SOURCES USED IN FORMULATING THE POSITIONS IN CHAPTER 6 AND VARIANT POSITIONS

Sources of the Four Pairs of Propositions

I offered the four pairs of propositions (section 6.2) in order to provide an interpretation of key points at issue in current controversies about TGs. I think (and this has been confirmed in numerous discussions) that they adequately express—not necessarily in a protagonist's own words, but in words designed to bring disagreements out sharply—positions held by important opposing groups, and that each group will readily accept that it holds the positions attributed to it (or positions very close to them). In the next section, I will list the principal sources of my statements of the propositions. Then, in the following section, I will outline a number of variant positions in the controversies. Articles representing the viewpoints of many (but not all) of the proponents and opponents that I identify can be found in the anthologies (Ruse and Castle 2002; Sherlock and Morrey 2002).

Pro

My statements of P_{1-4} have their origin in reading and talking with representatives of the main proponents of TGs. Without pretending to have achieved comprehensiveness of coverage, I cite representative sources of the various proponents. These include multinational agribusiness corporations (see quotations from Robert Shapiro, former CEO of Monsanto Corporation, in Specter 2000) and their clients, including many agronomists and farmers (see many quotations in Pollan 1998; Rauch 2003; and in other articles in magazines that aim for wide circulation), governments that endorse neoliberal policies and regard TGs to provide the key to export-oriented agricultural policies (Leite 2000; Araujo 2001), NGOs—for example, many organizations affiliated with CGIAR (Consultative

Group on International Agricultural Research)—that sponsor research intended to inform agriculture in ways that are in continuity with the green revolution (e.g., Persey and Lantin 2000; Pinstrup-Anderson and Schioler 2000; Borlaug 2000; 2003; NYT 2003b), a considerable body of molecular biologists and researchers in biotechnology (McGloughlin 1999; Pavan 2003; Prakash 2000), various commissions of experts (Nuffield Council on Bioethics 1999; Human Development Report 2001; NRC 1999; 2002), important parts of the scientific establishment (articles published in such journals as *Science*, *Nature*, *Nature Biotechnology*, and *AgBioForum* are generally, but not uniformly, favorable to posits like P_{1-4}; *AgBioView* is an almost daily newsletter that strongly favors the P-side), and many influential newspapers and magazines of opinion.

The U.S. trade lobby is also an important proponent. The United States has protested to the WTO against strong labeling regulations for TGs that have recently been introduced in the EU (Leonhardt 2003; Kindetlerer 2003; see Haslberger 2003 for some details of the regulations), replacing a five-year ban on planting TG crops and importing many of their products. The rhetoric of the protests notably reflects P_1 (and also upholds P_{2-4}) and the modern valuation of control. Supported by many biotech researchers and administrators (e.g., Ramón, MacCabe, and Gil 2004; Miller 2003), it has been objected that EU regulations are informed by claims ("fears") that are "unscientific" or that have no scientific support. For example, Reuters (June 19, 2003) quotes President Bush as telling a biotechnology conference: "Acting on unfounded, unscientific fears, many European governments have blocked the import of all new biotech crops. Because of these artificial obstacles, many African nations avoid investing in biotechnology, worried that their products will be shut out of important European markets." Then, he added, with allusions to P_4 (and P_3): "For the sake of a continent threatened by famine, I urge the European governments to end their opposition to biotechnology. We should encourage the spread of safe, effective biotechnology to win the fight against global hunger." (The Reuters article added that U.S. corn growers say that they are losing about $300 million per year because of the European trade barrier to TGs.) On the same day, the Associated Press reported a spokesman from the U.S. Trade Representative's office referring to the EU's "illegal and unscientific moratorium"; and the Speaker of the U.S. House of Representatives suggested that the moratorium was based on "fear rather than sound science" (Kindetlerer 2003). Also, a spokesman for the American Farm Bureau Federation added: "Countries shouldn't be able to erect barriers for non-scientific reasons. That's a very important principle in international trade" (Leonhardt 2003).

Con

In formulating C_{1-4} I have drawn on reading of and discussion with representatives and advisors of some of the Latin American movements—so called "popular organizations" (*SVF*: 185–86; Lacey 2003c) or, as some anthropologists call them, "new social movements" (Escobar 1995)—especially agriculturally oriented ones that have been participating in the World Social Forum (WSF). (Information and documents of WSF, most of them available in English translation, are available at www.forumsocialmundial.org.br (October 17, 2004); see also Loureiro, Cevasco, and Corrêa Leite 2002.) WSF meets at same time as the World Economic Forum (WEF) that usually meets in Davos, Switzerland, in late January or early February each year, to discuss themes of interest to the global-market ("neoliberal") system. To date WSF has organized, in addition to numerous regional meetings, four international meetings: 2001, 2002, and 2003 in Porto Alegre, Brazil; 2004 in Mumbai, India. WSF deliberately brings to the agenda of discussion issues that are sidelined at WEF, for example, those pertaining to the persistence and deepening of poverty that accompanies the spread of neoliberal policies. WSF is an international gathering place representing a great variety of self-characterized emancipatory or liberation movements, committed to the view that only out of their collaborative struggles can "another world," one that embodies social justice more fully, come into being (see chapter 11).

I have drawn most from reflections on the Brazilian movement, MST (Movimento dos trabalhadores rurais Sem Terra—Movement of Landless Rural Workers). On MST, see Stédile (1999); Branford and Rocha (2002); Wright and Wolford (2003). MST is in the forefront of resistance to the spread of TGs in Brazil (see MST/SP 2004). Its members are also involved in the growing popularity of agroecology. See (in addition to the works just cited), for example, "Carta da 3a Jornada de Agroecologia: construindo um projeto popular e soberano para a agricultura camponesa," June 2004, www.jornadadeagroecologia.com.br (September 3, 2004); and AS-PTA (2004). On issues connected with WSF and MST (and the role of agroecology in them), I have benefited greatly from collaboration with Universidade de São Paulo philosopher of science, Marcos Barbosa de Oliveira (see Lacey and Barbosa de Oliveira 2001), and from discussions with Universidade Federal de Santa Catarina, Florianópolis agronomists Miguel Guerra and Rubens Nodari (see Guerra et al. 1998a; 1998b; Nodari and Guerra 2000; 2001). C_{2-3} are familiar themes of many well-known international NGOs. C_2 is usually part of criticisms of neoliberal globalization policies and projects; Greenpeace and the Canadian based ETC (see bibliography) have been in the forefront of making the case for C_3 and publicizing it; and the California based Food First has assembled evidence in support of C_4—Food First: Institute for Food and Development Policy, www.foodfirst.org (October 17, 2004); see also Independent Science Panel (2003), and articles by Rosset (see bibliography), which offer a spirited case for C_3 and C_4.

C_1 is not so familiar. Agroecologists, such as Miguel Altieri, have conducted their research with the awareness that their methods (although they possess sound scientific credentials) lie outside of mainstream methods, and the success of their research shows that fruitful research under alternative strategies is possible (section 5.4 and section 10.1); but they have not systematically thematized how their approach both is complementary to and in competition with research conducted under materialist strategies. For a defence of C_1, in a context that shows awareness of the importance of C_4, see Costa Gomes and Rosenstein (2000): this work and Machado and Fernandes (2001) point to the productive potential that can be achieved following research on agroecological methods. Guerra et al. (1998a; 1998b) emphasize the possibilities of exploiting in farming the rich biodiversity of crop plants that is present in poor tropical countries. And propositions with close affinities to C_1 have been explicitly affirmed along with C_{2-4} (indeed C_1 serves to deepen the criticism of P_{2-4} represented by C_{2-4}) within WSF (Barbosa de Oliveira 2002; Berlan 2001; Lacey 2002d) and by well-known participants in WSF such as Vandana Shiva (section 7; see her works listed in bibliography), Mae-Wan Ho (1998; 2000) and Laymert Garcia dos Santos (2003), and P_1 (or views close to it) have been criticized (not upholding C_1 in all cases) in a recent volume organized by another well-known participant in WSF, Boaventura de Sousa Santos (Santos 2005). WSF also articulates the values of "popular participation" (see examples quoted in chapter 11).

Variant Positions

Pro

P_{1-4} represent my encapsulations of controversial posits that constitute the core of many arguments that aim to legitimate the current rapid implementation and increasingly widespread use of TG-oriented farming, and to influence the formation of public agricultural and trade policies. They serve to sharpen what is at stake in the controversies.

There is variety among the proponents of TGs, however; not all of them uphold all of the four stated propositions, and, in some cases, different interpretations of them are made. Concerning P_2, for example, there are those, often linked with CGIAR (see above), who put their authority behind the promises held out for the future and behind policies that emphasize research and development said to be aimed at serving the peoples of impoverished countries by, for example, producing golden rice (section 8.3); but who at the same time question the benefits of the currently most widely used TGs (having tolerance to proprietary herbicides or pesticide properties). In line with C_2, they question whether these are benefits for anyone who is not driven by the profit motive, since these TGs have been introduced not to address the needs of poor farming communities or for the sake of producing increased yields in non-optimal soils and of health-enhancing foods, but for the sake of increasing sales of particular pesticides or gaining greater control of the seed market (e.g., Nuffield Council on Bioethics 1999; Serageldin 1999). Then, concerning P_4, these people are likely to recognize a more modest (though still necessary) role for uses of TGs in the agriculture of the future than those (especially agribusiness) who uphold P_2 without qualification, and to be more open to cooperative ventures with those who hold C_4 (ABC 2003; FAO 2004). Concerning P_3 there are also differences of interpretation, reflecting different views about the appropriate standards to which empirical studies on risk should be held. Professional scientists, in particular, are generally likely to resist making unqualified generalizations about the risks of TGs and to insist that the potential risks of TGs need to be investigated variety by variety and environment by environment; they may then qualify P_3 to refer to specific varieties of TGs grown in specified environments (in which conformity with well-designed regulatory standards can be counted on), and point to the importance of on-going monitoring of TGs that have been released for use (NRC 2002).

I have indicated that P_{1-4} function as premises of arguments for the legitimacy and necessity of developments and uses of TGs. (The qualified versions of the previous paragraph suffice to serve this function.) There is another pro TG viewpoint that is worth considering. Paul Thompson elaborates it in detail and with nuance (Thompson 1997). Echoes of this position can also be heard sporadically throughout NRC (2002), a report of a commission on which Thompson served (see also section 10.6). It upholds P_{1-3} (perhaps with qualified interpretations of P_2 and/or P_3), and thus defends the legitimacy of planting TG crops at least in some environments. But it leaves open that C_4 may turn out to be true, and so (at least for the time being), it questions the value and the need for public agricultural policies that prioritize developments of TGs rather than of alternatives such as agroecology, especially in societies where there are social movements that contest the modern valuation of control.

This viewpoint has affinities with another (that I have heard in discussions with agronomists in Brazil). On the one hand, it is opposed to agribusiness, its objectives, and the difficulties—deriving from the fact that various genetic materials and techniques of genetic engineering cannot be freely obtained and used because of intellectual property protections—that it creates for research of the kinds of TGs in which they might be interested; it also sees the programs of CGIAR as "arrogant," attempting to produce a few varieties of seeds for use regardless of local ecological, climatic, social, and other conditions. But, on the other hand, it claims that there are special local problems (e.g., widespread fungus infections, whose spread is furthered by conditions of permanent humidity) that are not amenable to agroecological solution, but for which a solution could be found by developing appropriate varieties of TG crops. (This viewpoint might be held together with the further one that, although agroecological developments are needed to improve the food situation of the poor in rural areas, they cannot produce sufficient food to feed large concentrated urban populations.) Then, developments of TGs would be necessary, but now this claim must be appropriately qualified in light of recognizing that agroecology should play a role too; and the degree of prominence TGs should have in public agricultural policy becomes a matter for debate (see section 10.4).

Con

There is even more variety on the C-side than there is on the P-side. C_{1-4} are the foundation, in my opinion, of the most compelling line of argument against the legitimacy and necessity of using TGs at the present time—in virtue of their primary thrust being a positive one in favor of developments of such alternative forms of farming as agroecology (for clarification about "agroecology," see chapter 10: note 1) that offer themselves as better alternatives not only to TG-oriented agriculture, but also to conventional high intensity agriculture that is heavily dependent on the use of petrochemicals. Some critics make an essentially negative argument that involves only C_2 and especially C_3. Many have been preoccupied with the issue of risks (Lappé and Bailey 1998): risks to human health and the environment that may arise from some or other of the planting, production, processing, and consumption of TGs; the alleged inadequacy of current risk-assessment procedures and regulatory bodies (Pollack 2003; Bergelson and Purrington 2000); threats to biodiversity and the "pollution" of centers of biodiversity of crops such as maize (section 9.5); dangers of corporate control of the food supply; loss of local and regional food sovereignty (Rosset 2003) and thus loss of culturally valued food varieties; and potential undermining of the conditions for organic farming. Risks of these kinds can be investigated scientifically (Risler and Mellon 1996; Bergelson, Winterer, and Purrington 1999; Bergelson and Purrington 2000), although relevant investigations take a lot of time and raise difficult methodological issues (chapter 9).

Some who focus on risks are often convinced (like the CGIAR proponents) of the future promise of TGs, and they are convinced that, given time, the issues about risks can be satisfactorily resolved, that some varieties of TGs will pass rigorous risk assessments for use in specified environments with specified regulatory oversight in place. Thus, they may urge taking special precautionary measures, adopting the "Precautionary Principle" (section 9.4) (those who uphold the C-side, as presented in the text, may also adopt this principle), and also labeling of TG products, in line with claims about the right of consumers to know the ingredients of the food they buy and eat (Ruse and Castle 2002: articles in part IV). (Support for strong regulations about labeling is particularly widespread in Europe where, following recent problems with "mad cow disease," regulatory bodies are often treated with suspicion.)

Others maintain that these risks cannot or are unlikely to be adequately managed and their effects contained, and so that together they add up to a case for dropping the whole TG approach, and perhaps for engaging in active opposition.

Such active opposition may involve making legal challenges or engaging in consumer resistance to buying TG products. For example, one legal challenge, made to the European Patent Office, succeeded in getting nullified patent claims to *neem* products, granted to the U.S. Department of Agriculture and multinational W.R. Grace Corporation (Shiva 2000c). Shiva also filed suit in India's Supreme Court seeking to stop field trials of TG cotton on the ground that the safety of the trials has not been well established with available empirical data (Normile 2000). In appendix: 2, I discuss the action of a Brazilian court in staying the selling and planting of TG soybean, pending carrying out an environmental impact study. Active opposition may also lead to destroying TG crops that have been planted or disrupting the transportation of their products, and using tactics of civil disobedience—aiming to prevent further research, implementation, or distribution. For example, *O Estado de São Paulo* (November 8, 2000) reported that João Pedro Stédile (an MST leader) announced that MST "would burn all TG crops that it found, as well as impede the export of these products from Brazilian ports"; and also: " in yesterday's protest members of MST bombarded the principal entrance of the U.S. Embassy with ears of TG maize" (my translation). Then, a few months later, during the WSF meeting in Porto Alegre (January 25–30, 2001), Stédile, accompanied by the French activist Jose Bové organized the uprooting and subsequent burning of TG soybean (which Stédile claims to have been illegally grown) at a nearby Monsanto experimental farm. (In Brazil, it is not only protagonists of the C-side who have resorted to civil disobedience. In recent years, farmers in the state of Rio Grande do Sul have planted TG crops illegally, following smuggling the seeds into Brazil from Argentina.) Sometimes arguments based on risks are deepened by appeal to C_1 and/or C_4; at other times they are considered decisive by themselves.

There is another cohort of critics who reject outright, or harbor strong apprehensions, toward the "intrusion into nature" that planting TGs exemplifies. Often this position comes in religious garb. The most famous proponent has been the Prince of Wales and his remarks have been widely disseminated in the press, as in the following comment: "He [Prince Charles] used a nationally broadcast BBC lecture last month (Prince of Wales 2000; also 1998; cf. Görgen 2000) to raise the moral stakes in the debate, . . . saying that experimentation [on TGs] was taking man into realms that belong to 'God and God alone.' 'We need to rediscover a reverence for the natural world, irrespective of its usefulness to ourselves. If literally nothing is held sacred anymore, what is there to prevent us treating our whole world as some grand laboratory of life with potentially disastrous long-term consequences'" (Hoge 2000). Whether or not put in religious garb, this viewpoint is a descendant of the Romantic reaction to the Enlightenment. It has gained notoriety from the coining of the term "Frankenfoods" to refer to the products of TG crops. As it stands, this position lies outside of the scope of the argument framed by C_{1-4}, which by positing C_1 attempts to put the conflict between each of the other pairs of propositions into the arena where scientific investigation is possible. No doubt it would uphold modified versions of C_{1-4}, but the view represents a stance towards the world that is intended not to be vulnerable to the outcomes of scientific inquiry, so that endorsing C_1 is not fundamental to it. Here I only observe that this position is different from the one, which I discuss in the main text, in which reflection on scientific issues is central. I leave it an open question whether a version of the religious argument could be synthesized with the science-based argument, and whether there could be a place in science-based arguments, based on ecological considerations, for using such value-laden concepts as "intrusion into nature" (cf. Mariconda and Carvalho Ramos 2003). Sometimes the P-side claims that the C-side represents an "anti-science" ("luddite") stance (Borlaug 2000; Potrykus 2001). The "romantic" position perhaps does. Arguments based on C_{1-4} do not.

2. ADDITIONAL MATERIAL ON RISKS

Additional Discussion of Risks

For additional discussion of the kinds of risks listed in section 9.1, see (among many others) Ho (2000a); Nodari and Guerra (2001); Rissler and Mellon (1996); Altieri (2001); Leite (2000). Ho raises particular concerns about the possibility of "horizontal gene transfer" (of genes inserted into the genomes of crop plants or further recombination of them into plants or other organisms in the environment), that can lead to the generation of infectious diseases, that may be facilitated by genetic material from the Cauliflower Mosaic Virus being used as a "promotor" in many of the technologies of genetic modification (Ho, Ryan, and Cummins 1999). (I am unable to evaluate the evidence offered by Ho and her associates and the counterarguments of their critics.) Note that this virus is utilized in the engineering of "golden rice." Heinemann and Traavik (2004) discuss horizontal gene transfer in connection with "the massive release of antibiotics into the environment" that is occasioned by some current uses of TGs and methodological problems connected with detecting its frequency. Concerns about risks of contamination and the emergence of "superweeds" have delayed, pending more detailed environmental impact studies, the commercial release of TG grasses intended for use on golf courses and home lawns (Pollack 2004b). There has been a lot of discussion about the risk of creating "superweeds," but there don't seem to be any clear generalizations established. Some studies (backed by theoretical arguments) suggest that the emergence of certain kinds of superweeds is very unlikely (Adam 2003), others, that other kinds are not so unlikely (Dalton and Diego 2002). The researcher cited in the last report was denied access, by the companies that developed them, to both the transgene and the plant seeds for follow-up research.

Principle of Substantial Equivalence

When dealing with potential risks to human health, the P-side often appeals to the "principle of substantial equivalence." Simplifying, it proposes that the relevant proteins (those pertinent to how food is ingested into the body) produced by TG plants (e.g., of soybean) are the same (have the same chemical composition) as those produced by plants grown in conventional agriculture. Therefore, it is said, prima facie TG crops are as safe as conventional ones, and their products should not be held to need special testing—at least until such time as explicit evidence to the contrary is provided. The principle of substantial equivalence functions to put the burden of proof on the C-side. It is controversial and widely discussed, for example, in Millstone, Brunner, and Mayer (2002); Schauzu (2000); Lewontin (2001). The notion of "equivalence" involved is questioned in Mariconda and Carvalho Ramos (2003). This principle has no relevance to the discussion of environmental risks.

Recent Studies in the UK

Giles (2003) reports on extensive field-scale studies (carried out in UK) that show "that two genetically modified crops—spring oilseed rape and beet—are likely to have harmful effects on farmland biodiversity. Researchers say that the level of weeds, seeds and insects in fields of transgenic crops were lower than those in plots of conventional varieties, and that this could have a knock-on effect on the birds and small animals that feed on these populations." The studies, cited by Giles, have been contested. One critic suggested that a closer examination of the data might indicate that the cause of the observed reduction in biodiversity was not the TG herbicide-resistant crops, but pesticide-spraying regimes (Mitchell 2003); others put it, "weed populations are a result of the management strategy, not the GM status of a crop" (Chassy et al. 2003), and they also criticized the methodology of the studies (as did Andow 2003).

Burden of Proof: Legal and Legislative Actions in Brazil

It is not always well understood that the results of risk assessments do not travel well across regional and national boundaries. The issue is central to legal actions that have been taken concerning TGs in Brazil. In 1998, the regulatory body appointed by the Brazilian government, CTNBio (Comissão Técnica Nacional de Biossegurança), approved varieties of RoundUp Ready soybeans for commercial release. In doing so, it declared that planting them involves no unacceptable risks to the environment, but its judgment was based on risk assessments conducted in the United States. Critics objected that testing (relating to environmental risks) of these varieties had not been carried out in the context of Brazilian agroecosystems; further, they objected, the prior burden had not been assumed in Brazil by the producers and users of these TGs. Subsequently, a Federal Court (responding to lawsuits brought on behalf of a number of NGO critics) stayed the release of these varieties pending environmental impact studies being conducted in Brazil (Cezar 2003; Leite 2000; Marinho 2003). The court's judgment, thus, is correctly and fully in accordance with proper scientific standards.

As of the day of revising this paragraph (September 30, 2004), this court ruling still stands, but a currently pending bill in the federal legislature may lead to its being withdrawn. A draft of legislation planned to be voted on by the Brazilian senate in the near future (projeto de lei n° 2.401-A, de 2003) gives CTNBio authority to liberate TGs for commercial use, but the Ministry of the Environment also has the responsibility to ensure that adequate risk assessment is carried out in Brazilian ecosystems. (Following the widespread illegal growing of TG crops by certain groups of farmers, a series of presidential decrees in the last couple of years have permitted the growing of some TG crops in some regions of Brazil, pending the adoption of definitive legislation.) It was reported in Brazilian newspapers on September 17, 2004, that leaders of a major farmers' organization in southern Brazil had announced that—legally or not—they intend to go ahead with planting TG crops in the next growing season (beginning in October 2004). Even if plantings in accord with regulative procedures are risk free, there is plenty of reason to question the safety of unregulated plantings. For example, since the TG seeds planted have been smuggled into Brazil (mainly from Argentina) or saved from previous crops grown from smuggled seeds, there is no way to be confident that diseases (e.g., fungal, viral) have not been imported with the seeds.

Prior Burden of Proof

A wide range of critical opinions can be found about how adequately the P-side has carried the prior burden of proof. There are those who think that much more thorough risk assessments need to be conducted, but who expect that some TGs will pass the tests and that some may have already passed them (e.g., Bergelson and Purrington 2000; Rissler and Mellon 1996). Others question the adequacy of the temporal and spatial scales of standard risk assessments, whether in view of the studies about contamination or of theoretical reasons to doubt the long-term sustainability of TG-oriented (and conventional high-intensity) farming (Miguel Guerra: personal communication, March 2, 2004). The experiments described by Tilman (1998)—long-term, ten (or more)-year comparisons of high-intensity methods with "organic" alternatives showing them to have equivalent crop yields and comparable profitability, but where the latter "can improve soil fertility and have fewer detrimental effects on the environment"—provide an indication of ways in which standard risk assessments could be supplemented so that long-term effects could be addressed. Still others (e.g., Ho 2000a), drawing broadly upon holistic conceptions of nature and ecology, expect that new mechanisms of (potentially catastrophic) risk will be discovered, and they point to possible mechanisms that are not taken into account in standard risk assessments. Ho, Ryan, and Cummins (1999), for example, suggest that horizontal gene transfer needs further investigation. The P-side usually retorts that critics, like Ho, are simply grasping at straws, proposing complicated and expensive research projects simply on the remote chance that risks will be uncovered. Perhaps they are, but it is part of the middle way that these matters be stated explicitly and then argued out in public forums.

Public Policy: National Variations

I have presented the P-side as arguing for the legitimacy of going ahead with the project to transform agricultural practices so that TG-oriented farming becomes an integral (even predominant) part of the agriculture of the future. My statement of the middle ground assumes that novel TG innovations will be a regular part of this project. (The claims for the "humanitarian" value of the project depend on this becoming realized.) The middle ground will have to be staked out differently in different countries. In the United States, for example, TG-oriented farming is already widespread, especially the use of TGs with herbicide resistance and toxicity to certain insects. Whether or not the prior burden, concerning testing for environmental (and other) risks, was adequately carried prior to their introduction, their widespread use is a fait accompli. Thus, for critics of their use, the key question is no longer, "Is their release for commercial ends legitimate?" but "Are likely risks sufficiently serious so as to warrant the cessation of their use?" The C-side would have to assume a greater burden in addressing the latter question. Even so, the fact that there is question about the adequacy with which the prior burden was carried (e.g., monarch butterfly, contamination) reinforces the importance of it being carried adequately concerning any additional innovations. Compare (anticipating the next paragraph): "In promoting a more humble and less rapacious attitude to the environment, the precautionary principle presents a profound challenge to some of the unstated assumptions of modern (and particularly Western) societies: material growth, the power and efficacy of scientific reason, and the pre-eminence of human interests over those of other entities" (Jordan and O'Riordan 1999).

The Precautionary Principle

The precautionary principle does not have a settled formulation. Versions of it are acknowledged in some national laws (including in the draft Brazilian law on the regulation of TGs—see above) and international agreements. For example, Article 12, item 8 of the *Cartagena Protocol on Biodiversity* states: "Lack of scientific certainty due to insufficient relevant scientific information and knowledge regarding the extent of the potential adverse effects of a living modified organism on the conservation and sustainable use of biological diversity in the Party of import, taking also into account risks to human health, shall not prevent that Party from taking a decision, as appropriate, with regard to the import of that living modified organism intended for direct use as food or feed, or for processing, in order to avoid or minimize such potential adverse effects" (Cartagena Protocol 2002). There is tension between the precautionary approach, permitted under this Protocol, and the rules of WTO that permit trade restrictions "based only on concerns that are demonstrably scientific rather than those based on presumption or precaution as under the Cartagena Protocol" (Hodgson 2002b). The WTO rules acknowledge no prior burden to be carried by the P-side. So far as I know, the WTO has not yet ruled on complaints from the United States against EU restrictions on trade with TGs that appeal to the Cartagena Protocol. There is a vast literature exploring the implications of the precautionary principle and its various (not always consistent) formulations, much of it very critical. To get a good sense of the issues involved, see Raffensperger and Tickner (1999); Raffensperger and Barrett (2001); Goklany (2001); Morris (2000); Cezar and Abrantes (2003); Miller and Conko (2001).

Glossary of Abbreviations and Acronyms

AS	aim (or ideal) of scientific practices
B	a presupposition for identifying cv [in section 3.2]
C-side	the critics of using TGs
C_1–C_4	Propositions used by the C-side of the dispute about TGs
CGIAR	Consultative Group on International Agricultural Research [section 6: appendix]
cv	cognitive value (or, in context, cognitive values)
D	domain of phenomena (to which T applies)
EU	European Union
FTAA	Free Trade Association of the Americas
FS seeds	farmer-selected seeds
GMO	genetically modified organism (or TG organism)
IMF	International Monetary Fund
IPR	intellectual property rights
MR	methodological rule
MR-M	The MR: Entertain and pursue only theories that instantiate general principles of materialist metaphysics about the constitution and mode of operation of the world.
MR-OSV	The MR: Entertain and pursue only theories that are constrained so that, if accepted, they also manifest OSV.
MR-OVMC	The MR: Entertain and pursue only theories that are so constrained that, if soundly accepted, they also manifest OVMC.
MR-SM	The MR: Entertain and pursue *only* theories that are developed under S_M.
MR-SM'	The MR: Entertain and pursue theories that are developed under S_M.
MST	Movimento dos trabalhadores rurais Sem Terra—Brazilian Movement of Landless Rural Workers; an important member of FSM; involved with agro-ecological innovations.
NAFTA	North American Free Trade Association
NGO	non-governmental organization
NSE	No serious negative side effects; P_3 is an instance of NSE
NBW	No "better" way; P_4 is an instance of NBW
OSV	being an object of social value in the light of specified sv

274   Glossary of Abbreviations and Acronyms

OVMC	being an object of social value in the light of V_{MC} (or having significance in the light of V_{MC})
P-side	the proponents of using TGs
P_1–P_4	Propositions used by the P-side of the dispute about TGs
R&D	research and development
S	a strategy: constraints on theories and on the categories they may deploy/ identification of the kinds of possibilities to be investigated/ selection of the kinds of empirical data to be sought out and recorded
S_{AE}	agroecological strategies [defined in section 5.4]
S_{BT}	agro-biotechnological strategies; instances of S_M; the possibilities of transgenics are investigated under them.
S_{CSS}	strategies typically adopted in the contemporary social sciences
S_M	materialist strategies [defined in section 1.3]
S_{SS}	strategies for research in the social sciences, where adopting S_{SS} is linked with holding V_{PP}
SS	social structure
SS'	social structure that reflects emancipatory aspirations
SS_{NL}	structures of neoliberalism; structures that highly manifest $V_{C\&M}$
sv	a social value (or, in context, social values)
SVF	Is Science Value Free? Values and scientific understanding (Lacey 1999/2004)
T (T_1, T_2)	a theory (theories)
TG	transgenic
u	an object that has ø-value
v	a value, a characteristic that may be more or less manifested in some ø. [v may also be embodied in a social institution or structure: section 3.1.]
V	a value judgment concerning v. [Note the distinction between "value judgment" and "value-assessing statement": section 3.1.]
$V_{C\&M}$	values of capital and the market; values of neoliberalism [listed in section 6.3]; highly manifested in SS_{NL}
V_{MC}	the modern valuation of control [defined in section 1.1]
V_{PP}	values of "popular participation" [listed in section 6.3]
WSF	World Social Forum [appendix: 1; section 11]
WTO	World Trade Organization
X	a person (who holds values and adopts strategies)
ø-value	a value of ø; for example, an s-value (sv) is a value of social institutions and structures; cv and sv are instances of ø-values. [Note the distinction between "v is a ø-value" and "u is an object of ø-value": section 3.1.]

Bibliography

ABC—Academia Brasileira de Ciências. (2003). *Plantas Transgênicas na Agricultura.* Brazilia: Academia Brasileira de Ciências.
Adam, D. (2003). "Transgenic crop trial's gene flow turns weeds into wimps." *Nature* 421: 462.
AgBioView. (2004). An almost daily newsletter representing a strong pro-transgenics position, at www.agbioworld.org/newsletter_wm/index.php?caseid=archive (accessed October 5, 2004).
Albergoni, L., V. Pelaez, and M. Guerra. (2004). "Soja transgênica vs. soja convencional: Uma análise comparativa de custos e beneficios." Unpublished manuscript, Universidade Federal de Santa Catarina, Florianeopolis, Brazil.
Alcamo, I. E. (1996). *DNA Technology: The awesome skill.* Dubuque, Iowa: Wm. C. Brown Publishers.
Aldhous, P. (2000). "Genomics: Beyond the book of life." *Nature* 408: 894–96.
Alier, J. M. (2000). "International biopiracy versus the value of local knowledge." *Capitalism, Nature, Socialism: A journal of socialist ecology* 11 (June 2000): 59–68.
Altieri, M. A. (1987). *Agroecology: The scientific basis of alternative agricultures.* Boulder, Colo.: Westview.
———. (1994). *Biodiversity and Pest Management in Agroecosystems.* New York: The Haworth Press.
———. (1995). *Agroecology: The science of sustainable agriculture.* 2nd ed. Boulder, Colo.: Westview.
———. (1999). "The ecological role of biodiversity in agroecosystems." *Agriculture, Ecosystems and Environment* 74: 19–31.
———. (2000a). "No: Poor farmers won't reap the benefits." *Foreign Policy* 119: 123–27.
———. (2000b). "The ecological impacts of transgenic crops on agroecosystem health." *Ecosystem Health* 6: 13–23.
———. (2001). *Genetic Engineering in Agriculture: The myths, environmental risks, and alternatives.* Oakland, Calif.: Food First.
Altieri, M. A., and P. Rosset. (1999a). "Ten reasons why biotechnology will not help the developing world." *AgBioForum* 2: 155–62 (Reprinted in Sherlock and Morrey [2002].)
Altieri, M. A., and P. Rosset. (1999b). "Strengthening the case for why biotechnology will not help the developing world: A response to McGloughlin." *AgBioForum* 2: 226–36.

Altieri, M. A., P. Rosset, and C. I. Nicholls. (1997). "Biological control and agricultural modernization: Towards resolution of some contradictions." *Agriculture and Human Values* 14: 27–58.
Altieri, M. A., A. Yurjevic, J. M. Von der Weid, and J. Sanchez. (1996). "Applying agroecology to improve peasant farming systems in Latin America." In *Getting Down To Earth: Practical applications of ecological economics*, ed. R. Costanza, O. Segura, and J. Martinez-Alier. Washington, D.C.: Island Press.
Alves da Silva, J. M. (2000). "Os transgênicos e a sociedade rural." Op-Ed, *Folha de São Paulo*, September 18, 2000.
Anderson, E. S. (1995). "Knowledge, human interests, and objectivity in feminist epistemology." *Philosophical Topics* 23: 27–58.
———. (2002). "Situated knowledge and the interplay of value judgments and evidence in scientific inquiry." In *In the Scope of Logic, Methodology and Philosophy of Science*, vol. 2, ed. P. Gardenfors, K. Kijania-Placek, and J. Wolenski, 497–517. Dordrecht: Kluwer.
Andow, D. A. (2003). "UK farm-scale evaluations of transgenic herbicide-tolerant crops." *Nature Biotechnology* 21: 1453–54.
Andrén, O., H. Kirchmann, and O. Pettersson. (1999). "Reaping the benefits of cropping experiments." *Nature* 399: 14
Araujo, J. C. de. (2001). "Produtos transgênicos na agricultura—questões técnicas, ideológicos e politicos." *Cadernos de Ciência & Tecnologia* 18: 117–45.
———. (2003). "Transgênicos—Um olhar crítico sobre alguns mitos." *Cadernos ASLEGIS* 6 (21) (Dezembro): 1–12.
AS-PTA. (2004). Website of Assessoria e Serviços a Projetos em Agricultura Alternativa, Rio de Janeiro, at www.aspta.org.br/publique/cgi/cgilua.exe/sys/start.htm (accessed October 1, 2004).
Benbrook, C. (2003). "Sowing seeds of destruction." Op-Ed, *New York Times*, July 11, 2003.
Bergelson, J., and C. Purrington. (2000). "Factors affecting the spread of resistant *Arabidopsis thaliana* populations." In *Genetically Modified Organisms: Assessing environmental and human health effects*, ed. D. K. Letourneau and B. E. Burrows. Washington, D.C.: CRC Press.
Bergelson, J., J. Winterer, and C. B. Purrington. (1999). "Ecological impacts of transgenic crops." In *Applied Plant Biotechnology*, ed. V. L. Chopra, V. S. Malik, and S. R. Bhat. Enfield, N.H.: Science Publishers.
Berlan, J.-P. (2001). "Political economy of agricultural genetics." In *Thinking about Evolution: Historical, philosophical and political perspectives*, vol. 2, ed. R. A. Singh, C. B. Krimbas, D. B. Paul, and J. Beatty. Cambridge: Cambridge University Press.
Bhaskar, R. (1975). *A Realist Theory of Science*. Atlantic Highlands, N.J.: Humanities Press.
———. (1986). *Scientific Realism and Human Emancipation*. London: Verso.
———. (1993). *Dialectic: The pulse of freedom*. London: Verso.
———. (1998). *The Possibility of Naturalism*, 3rd ed. London: Routledge.
Borlaug, N. E. (2000). "Ending world hunger: The promise of biotechnology and the threat of antiscience zealotry." *Plant Physiology* 124: 487–90.
———. (2003). "The next green revolution." Op-Ed, *New York Times* July 11, 2003.
Boucher, D. H., ed. (1999). *The Paradox of Plenty: Hunger in a bountiful world*. Oakland, Calif.: Food First Books.
Branford, S., and J. Rocha. (2002). *Cutting the Wire: The story of the landless movement in Brazil*. London: Latin American Bureau.
Brush, S. B. and D. Stabinsky. (1996). *Valuing Local Knowledge: Indigenous people and intellectual property rights*. Washington, D.C.: Island Press.

Bunge, M. (1981). *Scientific Materialism*. Dordrecht: Reidel.
Carrier, M. (2001). "Changing laws and shifting concepts: On the nature and impact of incommensurability." In *Incommensurability and Related Matters*, ed. P. Hoyningen-Huene and H. Sankey. Dordrecht: Kluwer.
Cartagena Protocol. (2002). *Cartagena Protocol on Biodiversity of the Convention on Biological Diversity*, at www.biodiv.org/biosafety/protocol.asp?lg=1 (accessed October 2, 2004). The text of *The Convention on Biological Diversity*, adopted by many countries at the UN-sponsored Earth Summit (UN Conference on Environment and Development, Rio de Janeiro, 1992), is available at www.biodiv.org/convention/articles.asp (accessed October 2, 2004).
Cartwright, N. (1999). *The Dappled World: A study of the boundaries of science*. Cambridge: Cambridge University Press.
Cayford, J. (2003). "GMO opposition not based on a mistake." *Nature Biotechnology* 21: 493.
Cezar, F. G. (2003). *Previsões Sobre Tecnologias: Pressupostos epistemólogicos na análise de risco da soja transgênica*. Master's thesis, Universidade de Brasília.
Cezar, F. G., and P. C. Abrantes. (2003). "Princípio da precaução: Considerações epistemológicas sobre o princípio e sua relação com o processs de análise de risco." *Cadernos de Ciência & Tecnologia* 20: 225–62.
Chassy, B., C. Carter, M. McGloughlin, A. McHughen, W. Parrott, C. Preston, R. Roush, A. Shelton, and S. H. Strauss. (2003). "UK field-scale evaluations answer wrong questions." *Nature Biotechnology* 21: 1429–30.
Christou, P. (2002). "No credible scientific evidence is presented to support claims that transgenic DNA was introgressed into traditional maize landraces in Oaxaca, Mexico." *Transgenic Research* 11: iii–v.
Clarke, T. (2003). "Pest resistance feared as farmers flout rules." *Nature* 424:116.
Costa Gomes, J. C., and S. Rosenstein. (2000). "A geração do conhecimento na transição agroambiental: Em defesa da pluralidade epistemológica e metodológica na prática científica." *Cadernos de Ciência e Tecnologia* 17 (3): 29–57.
Coyne, J. A. (2003). "Doing acid." *New York Times Book Review*, June 15, 2003: 11–12.
Cupani, A. (1998). "A propósito do 'ethos' da ciência." *Episteme* 3: 16–38.
Dalton, R. (2003). "Traditional wheat breeder fights for funding." *Nature* 423: 790.
———. (2004). "Review of tenure refusal uncovers conflict of interest." *Nature* 430: 598.
Dalton, R., and S. Diego. (2002). "Superweed study falters as seed firms deny access to transgene." *Nature* 419: 655.
de Grassi, A. (2003). *Genetically Modified Crops and Sustainable Poverty Alleviation in Sub-Saharan Africa: An assessment of current evidence*. Third World Network–Africa, at www.eldis.org/static/DOC12623.htm (accessed September 27, 2004).
Doppelt, J. (2001). "Incommensurability and the normative foundations of scientific knowledge." In *Incommensurability and Related Matters*, ed. P. Hoyningen-Huene and H. Sankey. Dordrecht: Kluwer.
Douglas, H. (2000). "Inductive risk and values in science." *Philosophy of Science* 67: 559–79.
Dupré, J. (1993). *The Disorder of Things*. Cambridge, Mass.: Harvard University Press.
———. (2001). *Human Nature and the Limits of Science*. Oxford: Clarendon Press.
Echeverria, T. M. (2001). *Cenários do Amanhã: Sistemas de produção de soja e os transgênicos*. Doctoral thesis. Campinas: UNICAMP.
Ellacuría, I. (1991). "Utopia and prophecy in Latin America." In *Towards a Society That Serves Its People: The intellectual contribution of El Salvador's murdered Jesuits*, ed. J. Hassett and H. Lacey. Washington, D.C.: Georgetown University Press.

Escobar, A. (1995). *Encountering Development: The making and unmaking of the third world.* Princeton, N.J.: Princeton University Press.

ETC—Action Group on Erosion, Technology and Concentration. (2002). Communiqué, "Fear-reviewed science: The fight over Mexico's GM maize contamination." January 23, at www.etcgroup.org (accessed October 5, 2004).

FAO—Food and Agricultural Organization of the United Nations. (2004). *The State of Food and Agriculture, 2003-2004—Agricultural Biotechnology: Meeting the needs of the poor?* Rome: FAO, at www.fao.org/documents/show_cdr.asp?url_file=/docrep/006/Y5160E/Y5160E00.HTM (accessed October 5, 2004).

Feyerabend, P. K. (1981). *Problems of Empiricism.* Cambridge: Cambridge University Press.

——. (1993). *Against Method.* 3rd ed. London: Verso.

——. (1999). *Conquest of Abundance: A tale of abstraction versus the richness of being.* Chicago: University of Chicago Press.

Friedman, T. (2002). "Blunt questions, blunt answers." Op-Ed, *New York Times* Feb. 10, 2002.

Garcia dos Santos, L. (2003). *Politizar as Novas Tecnologias: O inpacto sócio-técnico da informacão digital e genetica.* São Paulo: Editora 34

Gewin, V. (2004). "Organic: Is it the future of farming?" *Nature* 428: 792–98.

Giere, R. N. (1999). *Science Without Laws.* Chicago: University of Chicago Press.

Giles, J. (2003). "Biosafety trials darken outlook for transgenic crops in Europe." *Nature* 425: 751.

Glanz, J. (2004). "Scientists say administration distorts facts." *New York Times*, Feb. 19, 2004.

Goklany, I. M. (2001). *The Precautionary Principle: A critical appraisal of environmental risk assessment.* Washington, D.C.: Cato Institute.

Görgen, S. A. (2000). "Transgênicos: Os riscos, o debate, a cautela necessária." In *Riscos dos Transgênicos*, ed. S. A. Görgen. Petrópolis, Brazil: Editora Vozes.

Gould, F. (2003). "*Bt*-management—theory meets management." *Nature Biotechnology* 21: 1450–51.

Guerinot, M. L. (2000). "The green revolution strikes gold." *Science* 287: 241–43. [Reprinted in Ruse and Castle 2002.]

Guerra, M. P., R. Nodari, M. S. dos Reis, and W. Schmidt. (1998a). "Agriculture, biodiversity and 'appropriate technologies' in Brazil." *Ciencîa e Cultura* 50: 408–16.

Guerra, M. P., R. Nodari, M. S. dos Reis, and A. I. Orth. (1998b). "A diversidade dos recursos genéticos vegetais e a nova pesquisa agrícola." *Ciência Rural, Santa Maria* 28: 521–28.

Hacking, I. (1999). *The Social Construction of What?* Cambridge, Mass.: Harvard University Press.

Haslberger, A. G. (2003). "Codex guidelines for GM foods include the analysis of unintended effects." *Nature* 21: 739–41.

Heinemann, J. A., and T. Traavik. (2004). "Problems in monitoring horizontal gene transfer in field trials of transgenic plants." *Nature Biotechnology* 22: 1105–09.

Hellmich, R. L., B. D. Siegfried, M. K. Sears, D. E. Stanley-Horn, M. J. Daniels, H. R. Mattila, T. Spencer, K. G. Bidne, and L. C. Lewis. (2001). "Monarch larvae sensitivity to Bacillus thuringiensis-purified proteins and pollen." *Proceedings of the National Academy of Science* 98: 11925–30.

Ho, M.-W. (1998). *The Rainbow and the Worm: The physics of organisms.* Singapore: World Scientific.

——. (2000a). *Genetic Engineering: Dream or nightmare.* New York: Continuum.

———. (2004a). "Modified seeds found among unmodified crops." *New York Times*, Feb. 24, 2004.

———. (2004b). "Genes from engineered grass spread for miles, study finds." *New York Times*, Sept. 21, 2004.

———. (2004c). "Can biotech crops be good neighbors?" *New York Times*, Sept. 26, 2004.

Pollan, M. (1998). "Playing God in the garden." *New York Times Magazine*, October 6, 1998.

Postlewait, A., D. D. Parker, and D. Zilberman. (1993). "The advent of biotechnology and technology transfer in agriculture." *Technological Forecasting and Social Change* 43: 271–87.

Potrykus, I. (2001). "Golden rice and beyond." *Plant Physiology* 125: 1157–61.

Powell, K. (2003). "Concerns over refuge size for US EPA-approved *Bt* corn." *Nature Biotechnology* 21: 467–68.

Prakash, C. S. (2000). "Can genetically engineered crops feed a hungry world?" at www.ipa.org.au/Units/Biotech/Prakashbiotech.html (accessed September 27, 2004).

Prince of Wales. (1998). "Seeds of disaster." *Ecologist* 28: 252–53.

———. (2000). "A royal view." BBC Reith Lectures 2000, reprinted in Ruse and Castle (2002).

Proctor, R. N. (1991). *Value-free Science? Purity and power in modern knowledge*. Cambridge, Mass.: Harvard University Press.

Putnam, H. (2002). *The Fact/Value Dichotomy and other essays*. Cambridge, Mass.: Harvard University Press.

Quist, D., and I. H. Chapela. (2001). "Transgenic DNA introgressed into traditional maize landraces in Oaxaca, Mexico." *Nature* 414: 541–43.

———. (2002). "Reply: Suspect evidence of transgenic contamination/Maize transgene results are artifacts." *Nature* 416: 602–3.

Raffensperger, C., and J. Tickner, eds. (1999). *Protecting Public Health and the Environment: Implementing the Precautionary Principle*. Washington, D.C.: Island Press.

Raffensperger, C., and K. Barrett. (2001). "In defense of the precautionary principle." *Nature Biotechnology* 19: 811–12.

Ramón, D., A. MacCabe, and J. V. Gil. (2004). "Questions linger over European GM food regulations." *Nature Biotechnology* 22: 149.

Rauch, J. (2003). "Will Frankenfood save the planet?" *Atlantic Monthly* 292 (October 2003): 103–8.

Rissler, J., and M. Mellon. (1996). *The Ecological Risks of Engineered Crops*. Cambridge, Mass.: MIT Press.

Rohatyn, F. (2002). "The betrayal of capitalism." *New York Review of Books*, Feb. 28, 2002, 8.

Rosset, P. (1999). "The multiple functions and benefits of small farm production." Policy Brief, no. 4, *Food First*, at www.foodfirst.org/pubs/policybs/pb4.html (accessed October 17, 2004).

———. (2001). "Genetic engineering of food crops for the Third World: An appropriate response to poverty, hunger and lagging productivity?" In *Proceedings of the International Conference on Sustainable Agriculture in the New Millenium—the Impact of Modern Biotechnology on Developing Countries*, at www.foodfirst.org/progs/global/biotech/belgium-gmo.html (accessed October 17, 2004).

———. (2003). "Food sovereignty: Global rallying cry of farmer movements." *Backgrounder* (Food First) 9 (4): 1–4.

Nagel, E. (1961). *The Structure of Science*. New York: Harcourt, Brace and World.
Nash, M. (2000). "Grains of hope: Genetically engineered crops could revolutionize farming. Protesters fear they could also destroy the ecosystem. You decide." *Time Magazine*, August 4, 2000.
Nature (2002). Editorial note, *Nature*: 416: 600.
Nature Biotechnology (2004). "Orphans at the window." Editorial, *Nature Biotechnology* 22: 1055.
Nodari, R. O., and M. P. Guerra. (2000). "Biossegurança de plantas transgênicas." In *Riscos dos Transgênicos*, ed. S. A. Görgen. Petrópolis, Brazil: Editora Vozes.
———. (2001). "Avaliação de riscos ambientais de plantas transgênicas." *Cadernos de Ciência e Technologia* 18: 61–116.
———. (2004). "Os impactos ambientais." *Ciência Hoje* 34 (203) (special issue on risks, benefits, and uncertainties of transgenics): 43–45.
Normile, D. (2000). "Agrobiotechnology: Asia gets a taste of genetic food fights." *Science* 289: 1279–81.
NRC—National Research Council. (1999). *Genetically Modified Pest-Protected Plants: Science and regulations*. A report of the Committee on Genetically Modified Pest-Protected Plants, Board on Agriculture and National Resources, National Research Council. Washington, D.C.: National Academy Press.
———. (2002). *Ecological Risks of Transgenic Crops: The scope and adequacy of regulation*. Washington, D.C.: National Academy of Sciences.
Nuffield Council on Bioethics. (1999). *Genetically Modified Crops: The social and ethical issues*. London: The Nuffield Foundation.
NYT—*New York Times*. (2003a). "Harvesting poverty: The rigged trade game." Editorial, July 20, 2003.
———. (2003b). "Genetically modified food and the poor." Editorial, October 13, 2003.
———. (2004a). "Science or politics at the F.D.A." Editorial, February 24, 2004.
———. (2004b). "The travels of a bioengineered gene." Editorial, September 30, 2004.
Oberhauser, K. S., M. Prysby, H. R. Mattila, D. E. Stanley-Horn, M. K. Sears, G. P. Dively, E. Olson, J. M. Pleasants, W.-K. F. Lam, and R. L. Hellmich. (2001). "Temporal and spatial overlap between monarch larvae and corn pollen." *Proceedings of the National Academy of Science* 98: 11913–18.
Pavan, C. (2003). "Está provado cientificamente: Os OGMs são mesmo seguros." Interview (June 4, 2003), *Conselho de Informações sobre Biotecnologia*, www.cib.org.br/entrevista.php?id=22 (accessed October 4, 2004).
Pennisi, E. (2000). "Breakthrough of the year. Genomics comes of age." *Science* 290: 2220–21.
Persey, G. J., and M. M. Lantin. (2000). *Agricultural Biotechnology and the Poor*. Washington, D.C.: CGIAR and U.S. National Academy of Science, at www.cgiar.org/biotech/rep0100/contents.htm (accessed October 16, 2004).
Pinstrup-Anderson, P., and E. Schioler. (2000). *Seeds of Contention: World hunger and the global controversy over GM crops*. Baltimore: The Johns Hopkins University Press.
Pleasants, J. M., R. L. Hellmich, G. P. Dively, M. K. Sears, D. E. Stanley-Horn, H. R. Mattila, J. E. Foster, P. L. Clark, and G. D. Jones. (2001). "Corn pollen deposition on milkweeds in and near cornfields." *Proceedings of the National Academy of Science* 98: 11919–24.
Pollack, A. (2000). "First Complete Plant Genetic Sequence Is Determined." *New York Times*, December, 14, 2000.
———. (2003). "Report says more farmers don't follow biotech rules." *New York Times*, June 19, 2003.

Loureiro, I. M., M. E. Cevasco, and J. Corrêa Leite, eds. (2002). *O Espírito do Porto Alegre*. São Paulo: Paz e Terra.

Louwaars, N. L., B. Visser, J.-P. Nap, and W. Brandenburg. (2002). "Transgenes in Mexican maize landraces, an analysis of data and potential impact." *Biotechnology and Development Monitor*, at www.biotech-monitor.nl/4910.htm (accessed September 2004).

Lutzenbeger, J., and E. Goldsmith. (2001). "Killing off small farms in Brazil." *Ecologist Report*, June 2001: 27–29.

Machado, A. T., and M. S. Fernandes. (2001). "Participatory maize breeding for low nitrogen tolerance," *Eupytica* 122: 567–73.

Machamer, P., and H. Douglas. (1999). "Cognitive and social values." *Science and Education* 8: 45–54.

Mariconda, P., and M. de Carvalho Ramos. (2003). "Transgênicos e ética: A ameaca à imparcialidade cientifica." *Scientiae Studia* 1: 225–46.

Mariconda, P., and H. Lacey. (2001). "A águia e os estorninhos: Galileo e a autonomia da ciência." *Tempo Social* 13 (1): 49–65.

Marinho, C. L. C. (2003). *O Discurso Polissêmico sobre Plantas Transgênicas no Brasil: Estado da arte*. Doctoral thesis, Escola Nacional de Saude Pública, Rio de Janeiro.

Martinez-Soriano, J. P., A. M. Bailey, J. Lara-Reyna, and D. S. Leal-Klevezas. (2002). "Transgenes in Mexican maize." *Nature Biotechnology* 20: 19–20.

Maxwell, N. (1984). *From Knowledge to Wisdom: A revolution in the aims and methods of science*. Oxford: Blackwell.

McGloughlin, M. (1999). "Ten reasons why biotechnology will be important to the developing world." *AgBioForum* 2: 163–74. (Reprinted in Sherlock and Morrey [2002].)

McMullin, E. (1983). "Values in science." In *PSA 1982*, vol. 2, ed. P. D. Asquith and T. Nickles. East Lansing, Mich.: Philosophy of Science Association.

———. (1999). "Materialist categories?" *Science and Education* 8: 36–41.

McNeil, D. G., Jr. (2004). "U.S. scientist tells of pressure to lift bans on food imports." *New York Times*, Feb. 25, 2004.

Mellon, M., and J. Rissler. (2004). *Gone to Seed: Transgenic contaminants in the traditional seed supply*. Washington, D.C.: Union of Concerned Scientists, at www.ucsusa.org/food_and_environment/biotechnology/page.cfm?pageID=1315 (accessed September 21, 2004).

Merton, R. (1957). *Social Theory and Social Structure*. Glencoe, Ill.: Free Press.

Metz, M., and J. Fütterer. (2002). "Suspect evidence of transgenic contamination." *Nature* 416: 600–601.

Miller, H. I. (2003). "First salvo in transatlantic food fight is far from last word." *Nature Biotechnology* 21: 737–38.

Miller, H. I., and G. Conko. (2001). "Precaution without principle." *Nature Biotechnology* 19: 302–3.

Milloy, S. (2004). "Notion of cow-human transmission is tenuous." *The Philadelphia Inquirer*, Commentary page, January 6, 2004.

Millstone, E., E. Brunner, and S. Mayer. (1999). "Beyond 'substantial equivalence.'" *Nature* 401: 525–26.

Mitchell, P. (2003). "Europe responds to UK's GM field trials." *Nature Biotechnology* 21: 1418–19.

Morris, J., ed. (2000). *Rethinking Risk and the Precautionary Principle*. Oxford: Butterworth, Heinemann.

MST/SP. (2004). *A Luta contra os Transgênicos: Subsídio para militância*. São Paulo: Vila Campesina Brasil.

———. (2005a). "On the interplay of the cognitive and the social in scientific practices." *Philosophy of Science* (forthcoming).

———. (2005b). "Science and human well-being: toward a new way of structuring scientific activity." In *Cognitive Justice in a Global World: Prudent knowledge for a decent life*, ed. B. de S. Santos. Madison: University of Wisconsin Press (forthcoming).

Lacey, H., and M. Barbosa de Oliveira. (2001). "Prefácio" a Vandana Shiva, *Biopirataria*. Petrópolis, Brazil: Editora Vozes. (Preface to the Portuguese translation of Shiva [1997a].)

Lacey, H., and B. Schwartz. (1986). "Behaviorism, intentionality and socio-historical structure." *Behaviorism* 14: 193–210.

———. (1987). "The explanatory power of radical behaviorism." In *B. F. Skinner: Consensus and controversy*, ed. S. Modgil and C. Modgil. London: Falmer Press.

———. (1996). "The formation and transformation of values." In *The Philosophy of Psychology*, ed. W. O'Donohue and R. Kitchener. London: Sage.

Lappé, M., and B. Bailey. (1998). *Against the Grain: Biotechnology and the corporate takeover of your food*. Monroe, Maine: Common Courage Press.

Laudan, L. (1977). *Progress and Its Problems: Towards a theory of scientific growth*. Berkeley: University of California Press.

———. (1984). *Science and Values: The aims of science and their role in scientific debate*. Berkeley: University of California Press.

Leisinger, K. M. (2000). "The 'political economy' of agricultural biotechnology for the developing world." Website of Novartis Foundation for Sustainable Development, at www.foundation.novartis.com/political_economy_agricultural_biotechnology.htm (accessed October 17, 2004).

Leite, M. (2000). *Os Alimentos Transgênicos*. São Paulo: PubliFolha.

———. (2003). "Futuros Xavantes." *Caderno Mais*, Sunday magazine of *Folha de São Paulo*, December 21, 2003.

Leonhardt, D. (2003). "Talks collapse on U.S. efforts to open Europe to biotech food." *New York Times*, June 20, 2003.

Lewanika, M. M. (2004). "GMOs and the food crisis in Zambia." In *The Gene Traders: Biotechnology, world trade, and the globalization of hunger*, ed. B. Tokar. Burlington, Vt.: Toward Freedom.

Lewontin, R. (1992). *Biology As Ideology: The doctrine of DNA*. New York: HarperCollins.

———. (1998). "The maturing of capitalist agriculture: Farmer as proletarian." *Monthly Review* 50 (3): 72–84.

———. (2000). *It Ain't Necessarily So: The dream of the human genome and other illusions*. New York: New York Review.

———. (2001). "Genes in the food!" *New York Review of Books* 48 (10): 81–84.

Lewontin, R., and J.-P. Berlan. (1990). "The political economy of agricultural research: The case of hybrid corn." In *Agroecology*, ed. C. R. Carroll, J. H. Vandemeer, and P. M. Rosset. New York: McGraw-Hill.

Longino, H. E. (1990). *Science as Social Knowledge*. Princeton, N.J.: Princeton University Press.

———. (1996). "Cognitive and non-cognitive values in science: Rethinking the dichotomy." In *Feminism, Science and the Philosophy of Science*, ed. L. H. Nelson and J. Nelson. Dordrecht: Kluwer.

———. (2002). *The Fate of Knowledge*. Princeton, N.J.: Princeton University Press.

Losey, J. E., L. S. Rayor, and M. E. Carter. (1999). "Transgenic pollen harms monarch larvae." *Nature* 399: 214.

———. (1977). "Objectivity, value judgment and theory choice." In *The Essential Tension*, ed. T. S. Kuhn. Chicago: University of Chicago Press.

———. (2000). "Commensurability, comparability, communicability." In *The Road Since Structure*, ed. T. S. Kuhn. Chicago: University of Chicago Press.

Lacey, H. (1985). "On liberation." *Cross Currents* 35: 219–41.

———. (1990). "Interpretation and theory in the natural and the human sciences: Comments on Kuhn and Taylor." *Journal for the Theory of Social Behavior* 20: 197–212.

———. (1997). "Neutrality in the social sciences." *Journal for the Theory of Social Behavior* 27: 213–41. (Reprinted in *Critical Realism: Essential readings*, ed. M. Archer, R. Bhaskar, A. Collier, T. Lawson, and A. Norrie. London: Routledge, 1998.)

———. (1998). *Valores e Atividade Científica*. São Paulo: Discurso Editorial.

———. (1999/2004). *Is Science Value Free? Values and Scientific Understanding*. London: Routledge. [Referred to throughout this book as *SVF*.]

———. (1999). "On cognitive and social values: A reply to my critics." *Science and Education* 8: 89–103.

———. (2000a). "Seeds and the knowledge they embody." *Peace Review* 12: 563–69.

———. (2000b). "Listening to the evidence: Service activity and understanding social phenomena." In *Beyond the Tower: Concepts and models for service learning in philosophy*, ed. C. D. Lisman and I. Harvey. Washington, D.C.: American Association for Higher Education.

———. (2001). "Incommensurability and 'multicultural science.'" In *Incommensurability and Related Matters*, ed. P. Hoyningen-Huene and H. Sankey, 225–39. Dordrecht: Kluwer.

———. (2002a). "The ways in which the sciences are and are not value free." In *In the Scope of Logic, Methodology and Philosophy of Science*, vol. 2, ed. P. Gardenfors, K. Kijania-Placek, and J. Wolenski, 513–26. Dordrecht: Kluwer.

———. (2002b). "Explanatory critique and emancipatory movements." *Journal of Critical Realism* 1: 7–31.

———. (2002c). "Assessing the value of transgenic crops." *Science and Engineering Ethics* 8: 497–511.

———. (2002d). "Tecnociência e os valores do Forum Social Mundial." In *O Espírito do Porto Alegre*, ed. I. M. Loureiro, M. E. Cevasco, and J. Corrêa Leite. São Paulo: Paz e Terra.

———. (2002e). "Where values interact with science." In *Siblings Under the Skin: Feminism, social justice and analytic philosophy*, ed. S. Clough. Aurora, Col.: The Davies Group Publishers, 2002.

———. (2003a). "Seeds and their socio-cultural nexus." In *Philosophical Explorations of Science, Technology and Diversity*, ed. S. Harding and R. Figueroa. New York: Routledge.

———. (2003b). "The behavioral scientist qua scientist makes value judgments." *Behavior and Philosophy* 31: 209–23.

———. (2003c). "Ellacuría on the dialectic of truth and justice." *Journal for Peace and Justice Studies* 13 (2003): 157–71.

———. (2003d). "The apt word." *Swarthmore College Bulletin*, September 2003: 20–21, at www.swarthmore.edu/news/commencement/2003/lacey.html (accessed September 15, 2004).

———. (2004a). "Assessing the environmental risks of transgenic crops." *Transformação* 27: 111–31.

———. (2004b). "Is there a significant distinction between cognitive and social values?" In *Science, Values and Objectivity*, ed. P. Machamer and G. Wolters. Pittsburgh: Pittsburgh University Press.

———. (2000b). "The 'golden rice'—an exercise in how not to do science." Website of Institute of Science in Society, at www.i-sis.org.uk/rice.shtml (accessed September 23, 2004).
———. (2002c). "FAQ on genetic engineering." Website of Institute of Science in Society, at www.i-sis.org.uk/FAQ.php (accessed October 16, 2004).
Ho, M.-W., A. Ryan, and J. Cummins. (1999). "Cauliflower mosaic viral promotor—a recipe for disaster." *Microbial Ecology in Health and Disease* 11: 194–97.
Hodgson, J. (2002a). "Doubts linger over Mexican corn analysis." *Nature Biotechnology* 20: 3–4.
———. (2002b). "UNEP [United Nations Environment Program] 'buys support for Cartegena' say critics." *Nature Biotechnology* 20: 205.
Hoge, W. (2000). "Britain's green prince and his family differ on altered crops." *New York Times*, June 7, 2000.
Horton, R. (2004). "The dawn of McScience." *New York Review of Books*, March 11, 2004.
Hoyningen-Huene, P. (1993). *Reconstructing Scientific Revolutions: Thomas S. Kuhn's philosophy of science*. Chicago: University of Chicago Press.
Human Development Report. (2001). *Making New Technologies Work for Human Development*. United Nations Development Programme, at www.undp.org/hdr2001/ (accessed October 17, 2004).
Independent Science Panel. (2003). *The Case for a GM-Free Sustainable World*, drafted by M.-W. Ho and L. L. Ching. London: Institute of Science in Society.
Jordan, A., and T. O'Riordan. (1999). "The precautionary principle in contemporary environmental policy and politics." In *Protecting Public Health and the Environment: Implementing the Precautionary Principle*, ed. C. Raffensperger and J. Tickner. Washington, D.C.: Island Press.
Kaplinsky, N., D. Braun, D. Lisch, A. Hay, and S. Hake. (2002). "Maize transgene results in Mexico are artifacts." *Nature* 416: 601–602.
Keller, E. F. (1996). "Feminism and science." In *Feminism and Science*, ed. E. F. Keller and H. E. Longino. Oxford: Oxford University Press.
Kindetlerer, J. (2003). "The WTO complaint—Why now?" *Nature Biotechnology* 21: 735–36.
Kitcher, P. (1993). *The Advancement of Science: Science without legend, objectivity without illusions*. New York: Oxford University Press.
———. (1997). *The Lives To Come: The genetic revolution and human possibilities*. New York: Touchstone.
———. (1998). "A plea for science studies." In *A House Built on Sand: Exposing postmodernist myths about science*, ed. N. Koertge. New York: Oxford University Press.
———. (2001). *Science, Truth and Democracy*. New York: Oxford University Press.
Kloppenburg, J., Jr. (1987). "The plant germplasm controversy." *Bioscience* 37: 190–98.
———. (1988). *First the Seed: The political economy of plant biology 1492–2000*. Cambridge: Cambridge University Press.
———. (1991). "Social theory and the de/reconstruction of agricultural science: Local knowledge for an alternative agriculture." *Rural Sociology* 56: 519–48.
Kloppenburg, J., Jr., and B. Burrows. (1996). "Biotechnology to the rescue? Twelve reasons why biotechnology is incompatible with sustainable agriculture." *Ecologist* 26: 61–67.
Knight, J. (2003). "Crop improvement: A dying breed." *Nature* 421: 568–70.
Kuhn, T. S. (1956). *The Copernican Revolution*. Cambridge, Mass.: Harvard University Press.
———. (1970). *The Structure of Scientific Revolutions*. 2nd ed. Chicago: University of Chicago Press.

Rouse, J. (1987). *Knowledge and Power: Towards a political philosophy of science*. Ithaca, N.Y.: Cornell University Press.
Rudner, R. (1953). "The scientist qua scientist makes value judgments." *Philosophy of Science* 20:1–6.
Ruse, M., and D. Castle, eds. (2002). *Genetically Modified Foods: Debating biotechnology*. Amherst, Mass.: Prometheus Books.
Sanchez, P. (2004). "The next Green Revolution." Op-Ed, *New York Times*, October 6, 2004.
Sankey, H. (1997). *Rationality, Relativism and Incommensurability*. Aldershot, U.K.: Ashgate.
Santis, S. de (2004). "Control through contamination: Genetically engineered corn and free trade in Latin America." In *The Gene Traders: Biotechnology, world trade, and the globalization of hunger*, ed. B. Tokar. Burlington, Vt.: Toward Freedom.
Santos, B. de S. (2002). *Reinventing Social Emancipation*, at www.ces.fe.uc.pt/emancipa/en/index.html (accessed July 31, 2004.).
———, ed. (2005). *Cognitive Justice in a Global World: Prudent knowledge for a decent life*. Madison: University of Wisconsin Press (forthcoming).
Schapiro, M. (2002). "Sowing disaster? How genetically engineered American corn has altered the global landscape." *Nation*, October 28, 2002.
Schauzu, M. (2000). "The concept of substantial equivalence in safety assessment of foods derived from genetically modified organisms." *AgBiotechNet* 2, at binas.unido.org/binas/reviews/Schauzu.pdf (accessed October 6, 2004).
Scriven, M. (1974). "The exact role of value judgments in science." In *Proceedings of the 1972 Biennial Meeting of the Philosophy of Science Association*, ed. R. S. Cohen and K. Schaffner. Dordrecht: Reidel.
Sears, M. K., R. L. Hellmich, D. E. Stanley-Horn, K. S. Oberhauser, J. M. Pleasants, H. R. Mattila, B. D. Siegfried, and G. P. Dively. (2001). "Impact of *Bt* corn pollen on monarch butterfly populations: A risk assessment." *Proceedings of the National Academy of Science* 98: 11937–42.
Serageldin, I. (1999). "Biotechnology and food security in the 21st century." *Science* 285: 387–89.
Sherlock, R., and J. D. Morrey, eds. (2002). *Critical Issues in Biotechnology*. Lanham, Md.: Rowman and Littlefield.
Shiva, V. (1988). "Reductionist science as epistemological violence." In *Science, Hegemony and Violence*, ed. A. Nandy. New Delhi: Oxford University Press.
———. (1991). *The Violence of the Green Revolution*. London: Zed.
———. (1993). *Monocultures of the Mind: Perspectives on biodiversity and biotechnology*. London: Zed Books.
———. (1997a). *Biopiracy: The plunder of nature and knowledge*. Boston: South End Press.
———. (1997b). Reply to Norman Borlaug's "Factual errors and misinformation." *Ecologist* 27: 211–12.
———. (1999). *Betting on Biodiversity: Why genetic engineering will not feed the hungry or save the planet*. New Delhi: Research Foundation for Science, Technology and Ecology.
———. (2000a). *Stolen Harvest: The hijacking of the global food supply*. Boston: South End Press.
———. (2000b). "Genetically engineered vitamin A rice: A blind approach to blindness prevention," at www.biotech-info.net/blind_rice.html (accessed October 16, 2004). (Reprinted in Ruse and Castle 2002.)
———. (2000c). "North-South conflicts in intellectual property rights." *Peace Review* 12:

Siegel, H. (2001). "Incommensurability, rationality and relativism." In *Incommensurability and Related Matters*, ed. P. Hoyningen-Huene and H. Sankey. Dordrecht: Kluwer.
Smith, J. E. (1996). *Biotechnology*. 3rd ed. Cambridge: Cambridge University Press.
Specter, M. (2000). "The pharmageddon riddle." *New Yorker*, April 10, 2000: 58–71.
Stanley-Horn, D. E., G. P. Dively, R. L. Hellmich, H. R. Mattila, M. K. Sears, R. Rose, L. C. H. Jesse, J. F. Losey, J. J. Obrycki, and L. Lewis. (2001). "Assessing the impact of Cry1Ab-expressing corn pollen on monarch butterfly larvae in field studies." *Proceedings of the National Academy of Science* 98: 11931–36.
Staples, B. (2004). "The battle against junk mail and spyware on the web." Editorial page, *New York Times*, January 3, 2004.
Stédile, J. P. (1999). *Questão Agrária no Brasil*. 7th ed. São Paulo: Atual Editora.
———. (2004). Address to the XII National Meeting of the MST (*Movimento dos trabalhadores rurias Sem Terra*—Movement of the Landless Rural Workers, January 2004, *MST Update* no. 55, published by Friends of MST, available from dawn@mstbrazil.org.
Taylor, C. (1982). "Rationality." In *Rationality and Relativism*, ed. M. Hollis and S. Lukes. Cambridge, Mass.: MIT Press.
Thompson, P. B. (1995). *The Spirit of the Soil: Agriculture and environmental ethics*. London: Routledge.
———. (1997). *Food Biotechnology in Ethical Perspective*. London: Blackie Academic and Professional.
———. (2003a). "The environmental case for crop biotechnology: Putting science back into environmental practice." In *Moral and Political Reasoning in Environmental Practice*, ed. A. Light and A. De-Shalit. Cambridge, Mass.: MIT Press.
———. (2003b). "Value judgments and risk comparisons: The case of genetically engineered crops." *Plant Physiology* 132: 10–16.
Tiles, M. (1996). "A science of Mars or Venus?" In *Feminism and Science*, ed. E. F. Keller and H. E. Longino. Oxford: Oxford University Press.
Tilman, D. (1998). "The greening of the green revolution." *Nature* 396: 211–12.
———. (1999). Reply to Andrén et al. (1999). *Nature* 399: 14
———. (2000). "Causes, consequences and ethics of biodiversity." *Nature* 405: 208–11.
Tilman, D., K. G. Cassman, P. A. Matson, A. Naylor, and S. Polasky, S. (2002). "Agricultural sustainability and intensive production practices." *Nature* 406: 719–22.
Tokar, B. (2004). *The Gene Traders: Biotechnology, world trade, and the globalization of hunger*. Burlington, Vt.: Toward Freedom.
Veblen, T. (1919). *The Place of Science in Modern Civilization and other essays*. New York: B.W. Huebsch.
Wambugu, F. (1999). "Why Africa needs agricultural biotech: There is an urgent need for the development and use of agricultural biotechnology in Africa to help counter famine, environmental degradation and poverty." *Nature* 400: 15–17.
Weiner, J. (1990). "Plant population ecology in agriculture." In *Agroecology*, ed. C. R. Carroll, J. H. Vandemeer, and P. M. Rosset. New York: McGraw-Hill.
Welsh, R., B. Hubbell, D. E. Ervin, and M. Jahn. (2002). "GM crops and the pesticide paradigm." *Nature Biotechnology* 20: 548–49.
Williams, B. (1985). *Ethics and the Limits of Philosophy*. Cambridge, Mass.: Harvard University Press.
Wittgenstein, L. (1958). *Philosophical Investigations*. Oxford, U.K.: Blackwell.
Wolfe, M. S. (2000). "Crop strength through diversity." *Nature* 406: 681–82.

Worthy, K., R. C. Strohman, and P. R. Billings. (2002). "Conflicts around a study of Mexican crops." *Nature* 417: 897–98.

Wright, A., and W. Wolford. (2003). *To Inherit the Earth: The landless movement and the struggle for the new Brazil.* Oakland, Calif.: Food First Books.

WSF—World Social Forum. (2001). "World Social Forum charter of principles," at www.forumsocialmundial.org.br/eng/gcartas.asp (accessed JJanuary 15, 2003).

———. (2002a). "Call of social movements—Resistance to neoliberalism, war and militarism: For peace and social justice," at www.forumsocialmundial.org.br/eng/portoalegrefinal_english.asp (accessed January 15, 2003).

———. (2002b). "Declaration of a group of intellectuals in Porto Alegre," at www.forumsocialmundial.org.br/eng/temas_declara_intele_POA_eng.asp (accessed January 15, 2003).

Ye, X., S. Al-Babili, A. Klöti, J. Zhang, P. Lucca, P. Beyer, and I. Potrykus. (2000). "Engineering provitamin A (Beta-carotene) biosynthetic pathway into (carotenoid-free) rice endosperm." *Science* 287: 303–5.

Zangerl, A. R., D. McKenna, C. L. Wraight, M. Carroll, P. Ficarello, R. Warner, and M. R. Berenbaum. (2001). "Effects of exposure to event 176 Bacillus thuringiensis corn pollen on monarch and black swallowtail caterpillars under field conditions." *Proceedings of the National Academy of Science* 98: 11908–12.

Zhao, J.-Z., J. Cao, Y. Li, H. Collins, R. Roush, E. D. Earle, and A. M. Shelton. (2003). "Transgenic plants expressing two *Bacillus thuringiensis* toxins delay insect resistance evolution." *Nature Biotechnology* 21: 1493–97.

Zhu, Y., H. Chen, J. Fan, Y. Wang, Y. Li, S. Yang, L. Hu, T. W. Mew, P. S. Teng, Z. Wang, and C. C. Mundt. (2000). "Genetic diversity and disease control in rice." *Nature* 406: 718–22.